T0345187

Unstable Nature

Unstable Nature is a popular science book offering a journey through the concept of instability in modern science with a focus on physics. Conceived for the curious reader wishing to go deeper in the fascinating and not yet popularised world of instabilities, it provides an immersion into paradoxical and unexpected phenomena – some of which hides in plain sight in our daily lives.

The book is written without technical jargon, and new concepts and terminology needed for the narrative are introduced gradually based on examples taken from accessible everyday life. The chapters are connected through a path that starts from exploring instabilities at the planetary scale and then passes through a description of unstable dynamics in macroscopic settings such as in human mechanical artifacts, fluid waves, animal skin, vegetation structures, and chemical reactions, finally reaching the sub atomic scale and the biological processes of human thought. Before concluding with some general philosophical remarks, a modern landscape about the possibility of seeing instabilities not only as a detrimental effect but as resources to be harnessed for technology is explored.

The book is enriched by a variety of professional anecdotes stemming from the direct research experience of the author. It features numerous connections of scientific concepts presented with other branches of the human experience and knowledge including philosophy, engineering, history of science, biology, chemistry, mathematics and computer science, poetry, and meditation.

Key Features:

- Presents an exciting introduction to the topic, which is accessible to those without a scientific background
- Explores milestone discoveries in the history of the concept of instability in physics
- Contains anecdotes of key figures from the field, including James C. Maxwell, Alan Turing, Vladimir Zakharov, Edward Lorenz, Enrico Fermi, and Mary Tsingou

Dr. Auro Michele Perego received a BSc and an MSc in Physics from Università degli Studi dell'Insubria (Como, Italy) and a PhD degree in Electrical Engineering – as a Marie Curie Early Stage Researcher – from Aston University (Birmingham, UK). He is currently a Royal Academy of Engineering Research Fellow at the Aston Institute of Photonic Technologies, where he leads his independent research group. His main research interests are: nonlinear science and the theory and applications of instabilities and self-organisation processes, fibre optics, laser physics, and photonic technologies such as optical frequency combs, optical amplifiers, optical sensing, and communication systems.

Unstable Nature

Order, Entropy, Becoming

Auro Michele Perego

CRC Press
Taylor & Francis Group
Boca Raton London New York

CRC Press is an imprint of the
Taylor & Francis Group, an **informa** business

First edition published 2024
by CRC Press
2385 NW Executive Center Drive, Suite 320, Boca Raton FL 33431

and by CRC Press
4 Park Square, Milton Park, Abingdon, Oxon, OX14 4RN

CRC Press is an imprint of Taylor & Francis Group, LLC

© 2024 Auro Michele Perego

ISBN: 9781032599601 (hbk)
ISBN: 9781032610917 (pbk)
ISBN: 9781003462040 (ebk)

DOI: 10.1201/9781003462040

Typeset in Minion
by Deanta Global Publishing Services, Chennai, India

Everything is so unstable, everything is so swirling, everything is in movement. There is no stability.

ATTRIBUTED TO ZYGMUNT BAUMAN

The popular scientific books by our scientists aren't the outcome of hard work, but are written when they are resting on their laurels.

LUDWIG WITTGENSTEIN

For Alba Fiamma

Contents

Acknowledgements

This book would have probably never been written if it were not for a serendipitous coincidence, as a consequence of which, in October 2021, Davide Pairone proposed me to write a broad scope scientific outreach work for the Italian publisher Aguaplano. I sincerely thank Davide Pairone and Raffaele Marciano of Aguaplano for believing in the project, and for skillfully guiding and advising me, with passion and care.

I would like to thank Rebecca Hodges-Davies, Danny Kielty, and the CRC Press team for their kind, careful, and passionate support, and for making the publication of *Unstable Nature* in the English language possible.

I am very grateful to my friend Greg Mathers for the superb and thorough copy editing work performed on my English translation of the original Italian version of the book.

My deepest gratitude for the time, for the patience, for the candour, goes to everyone who took care of reading – also more than once – *Unstable Nature* drafts, and of suggesting improvements, of individuating errors and imprecisions, of stimulating additions and clarifications: my wife Matilde Aliffi, my friends and scientist colleagues Tiziano D'Albis and Giulio Tirabassi, Professor Franco Prati, and Professor Stefano Trillo. Thanks to them the quality of the work has increased substantially; of the shortcomings, I am the only one responsible.

I wish to thank Professor Franco Prati for asking me, in a grey morning many years ago, while we were walking from the parking lot towards the university, which plans I had for my Masters dissertation. That one was a decisive bifurcation for me towards the world of instability.

It is for me a cause of great honour that Professor Stefano Trillo agreed to write the preface of this book. I am deeply grateful for it.

Unstable Nature has been written, in its essential parts, during the three months preceeding and during the three months immediately following my daughter Alba Fiamma's birth. With all my being I hope she could navigate with clear mind and fearless heart the turbulent sea of Life's instabilities. To her these pages are dedicated.

Auro Michele Perego
29th May 2023, Birmingham

Preface

THE IMPOSSIBILITY OF KEEPING a pencil in vertical equilibrium on its tip – or analogously a marble located on top of a cupola – is a simple manifestation of the *instability* phenomenon belonging to the experience of every human being. However, the concept of instability, in short the departure from equilibrium induced by perturbations both macroscopic and invisible to the human eye, is a concept which overlooks these simple examples and that, conversely, has hugely evolved during the historical course of human thought. In this evolutionary path lasting centuries, such a concept has become a sort of key for interpreting numerous natural phenomena involving spatial and temporal scales so deeply different to each other – from those characteristic of subatomic world to cosmological scales – which is even difficult to imagine without feeling quivers. It ended up transcending the world of physics, to land on the ones of chemistry, biology, and neuroscience, or of economics, spreading deep philosophical implications on the way. Among them, suffice it to know that instabilities are the seed leading to the discovery of deterministic chaos, namely the abrupt awakening from the mechanistic dream of the possibility to describing exactly the evolution of a systems starting from the knowledge of its initial state.

The author of this book, Auro Perego, takes us by the hand and leads us on a fascinating journey uncovering the very rich world of instabilities, ranging from the rediscovery of true milestones of both the near and remote past of scientific thought to extremely current scientific topics whose history is still to be written, the whole spiced up with numerous personal and historical anecdotes keeping the reader's curiosity constantly alive. In light of my experience (already more than thirty years, sic!) devoted to the study of instability phenomena first in optics and then in fluid dynamics always with technical instruments distinctive of hard sciences, it surprises me in an extremely positive way Auro's very natural ability to abandon the *comfort zone* of the specialist ("the unreasonable effectiveness of mathematics in the natural sciences" to quote the Nobel Prize laureate Eugene Wigner), identifying with a dimension where the description and the analysis of the phenomena is performed in name of simplicity, without the help of any mathematical tool, hence becoming accessible to everyone. Even more rare is the fact that the proposition of such a path comes from a young scientist in a phase of his scientific career which the predominant (and somehow deviant, in my opinion) culture of *"publish or perish"* would like it to be devoted to obtaining results which are comprehensible to specialists alone. Conversely, the vastness of interests and the eclecticism of the author has led to a very personal scientific and philosophical *story telling* where the narrative of numerous engaging facts from different origins to each other is made following a compass whose needle is constantly oriented on instabilities in nature. The result has been for me a charming reading, which I am sure will thrill many other readers too.

Stefano Trillo, Professor of Electromagnetic Fields at
Università di Ferrara

CHAPTER 1

Introduction

W HAT DO THE ROGUE waves that threaten to destroy ships on the ocean, Saturn's rings, research into the best way of increasing the amount of information carried via the internet, the vortexes which form in airplane turbines, optical clocks, the regular stripes that appear on the backs of certain animals, the working of lasers, a neuron's transmission of electrical impulses, chaos theory, and the disintegration of an atomic nucleus have in common? This book tries to articulate an answer to this apparently bizarre question. In short though, the answer is instability.

On the face of it, this answer may seem unsatisfying, but my hope is that by the end of these pages, its flavour will be fuller and more fulfilling, like a mountain landscape admired at the end of an exciting and slightly tiring ascent and not simply seen in a photograph. I hope to accompany you along a path which unravels on different levels and dimensions. The first dimension twists and turns through human history and goes from the dawn of modern physics to the present day, overlooking the future. But this path develops on an incomparably wider scale, for it also goes from what is much greater than the human being – from cosmic and planetary structures falling under the domain of celestial mechanics – to what occupies an infinitely smaller place

DOI: 10.1201/9781003462040-1

– like a single atom – requiring for its understanding the tools of quantum mechanics. It is a path which, like a two-faced Janus, reciprocally connects destruction with reconstruction, order with disorder, and the curiosity of idle contemplation with maximal practical utility. Our journey bridges different disciplines which may seem far from the daily life of many of us: physics, chemistry, biology, engineering, mathematics, and philosophy; but at the same time passes closer to us than what we believe, and maybe ultimately it touches us all, as human beings. It meets the life of some of the personalities which have left important traces on the walk of humanity and also my personal, modest, and limited research experience in the field of photonics, laser physics, and nonlinear systems, the prism through which the information and the stories I would like to talk about are filtered and refracted.

If you have the patience to follow me, I will show you how the concept of instability, as it is understood by modern science, manifests itself in light waves and waves which propagate in fluids like water, inside an atom, in the human brain, in the network through which electrical energy is distributed to citizens of entire nations, and in many other forms too. We will see that instability, despite being a destructive process connected to chaos, can also be the germ of new order, and how, if wisely harnessed, it can be a powerful resource in the development of technological devices. In the following pages I do my best to emphasise the importance of the concept of instability to the sum of human knowledge, but also its practical utility and its exciting beauty.

I would not be too surprised if a professional scientist who came across this book should find concepts which are expressed in a way which, in their view, is too pictorial or metaphorical, or if they felt disappointed that many important examples of instability have not been treated properly, or even mentioned at all. Simplification and the use of a non-rigorous language *must* be adopted in order to explain technical and sophisticated topics in a way that is accessible to everyone. Furthermore, obviously for reasons of space, only some examples can be introduced here. Those who possess the necessary tools to understand technical texts and scientific publications could definitely find greater satisfaction and pleasure in reading the latter. In order not to overload the narrative, bibliographic references and notes have been kept to a minimum. However, a brief series of more in-depth reference readings is provided at the end of the book.

Mathematical sciences are only one of many possible modes that can be used to describe the world and the human being. They generally develop in environments which are closed and isolated from wider society; however, their impact is decisive in both shaping the world around us and, indirectly, the minds of those who do not practice them as their occupation.

I will be happy if this text will let a bit of what scientific minds have developed, and promptly continue developing, sheltered by their academic refuges, seep into the general public, and to communicate the message that, beyond historical, intellectual, and practical fascination, the concept of instability is namely a tool which is available also to the thought of those who are not scientists. I wish that the concept of instability could constitute, for all those who desire it, an active, creative, and conscious way of looking at the rich, shimmering world of nature and of human soul and that it could be one – amongst others – interpretative key helping at unravelling the complexity of our tangled life and at throwing a faint light on the majestic and often obscure and unpredictable stride of the cosmos.

CHAPTER 2

Preludium in Ordinary Language

The limits of my language mean the limits of my world.

LUDWIG WITTGENSTEIN

Before exploring the concept of instability in modern science it makes sense to establish a starting point that is familiar to everyone, namely our ordinary and everyday language. We often use the words "stability" and "instability" in our everyday lives. One only has to read newspapers or follow political debate on television to bump into sentences like: "military intervention could disrupt equilibrium in the region and create instability" or "The absence of a clear majority at next election could undermine the whole country's stability". Likewise, we often use the words "stability" and "instability" when we converse with our friends. Who hasn't said or heard things like: "Paul is in a stable relationship with Eleanor", "All this uncertainty about the future makes me doubt its stability", "We need economic stability before we start a family", or "He stood, stable like an oak"?

In common language the word "stability" is used to describe the situation where objects or states of affairs remain unchanged

 DOI: 10.1201/9781003462040-2

or are in some sense constant in time. When we say that the rock that stands in our path is stable, we mean that it won't move; that if we climb on it with all our weight, it won't change its position or fall apart. In one sense we use the word "stability" to indicate the static nature of a thing or its robustness with respect to possible external influences, but we can also speak of the stability of a process or of a set of dynamics. We consider a process to be stable when it continues to happen largely undisturbed by interference from factors internal or external to the system itself. When we say that a person is mentally stable, we certainly do not want to say that no thoughts or emotions appear in their mind and that nothing changes; on the contrary, we mean that that person possesses an inner life where the flux of thoughts, emotions, and images develop in a dynamic and harmonious way that is well coordinated with their actions. Likewise, we say that a person's cardiac system is stable not when the heart of the subject has stopped beating, but when their heartbeat is regular, and systole and diastole follow one another without impediments, without any slackening or sudden and unexpected acceleration. When we talk about financial stability, we certainly do not refer to a state of stasis where nothing happens. What we mean is a situation where commercial and financial exchanges can happen continuously without impediment, and without any substantial and sudden changes in price. When we talk about stable processes, we are not referring to the structurally static nature of the system we are considering, but we do mean that there is something about it that remains constant and immutable. We are saying that a particular activity, a flux, an exchange of things, of ideas, or of substances *continues* to happen in a more or less regular way, and in line with certain expectations. It is to this continuity that we intuitively refer when we talk about the stability of a process.

The static nature or robustness of a thing, or the continuity of a process, define a sort of *status quo*. In our linguistic experience we have the intuition that instability presents a danger to the *status quo*; it is something which puts its existence at risk by threatening the robustness of a thing or the continuity of a process. The idea of instability is closely connected to the ideas of uncertainty and unpredictability, so the word "instability" is frequently used to describe social or political situations where big changes, that might be capable of subverting the existing order, could happen at any moment. For example, a political power vacuum could leave open the possibility of revolutionary groups taking control of a country. We talk about mental instability when a person seems to repeatedly act in an unpredictable and disproportionate way in situations where we would expect to see constant and predictable behaviour which aligns to a particular scheme that is generally accepted and which society considers valid.

Common language is rich in nuance and contradiction; words can have multiple meanings and subtle connotations. The meanings of words are frequently ambiguous, and this allows different interpretations whose fields of meaning partially overlap. However, this richness and variety does not prevent us from communicating in a more than satisfying way in daily lives, and it enables the realisation of some of the highest and most quintessentially human expressions like poetry, and literature. In science, there is an effort to define the meaning of each term in a clear way. Although the literature of various scientific disciplines abounds with expressions and words borrowed from common language, in general it aims to make more rigorous use of language. With different degrees of intensity depending on the particular discipline, scientific literature formulates thoughts and statements through mathematical formalism using equations and particular symbols.

Without a sharpening or a radical modification of language, certain features of the world could not be pinpointed at all. As different scientific disciplines advance, they actively mould and forge language to describe the phenomena that they encounter, and over time each creates *its own* language. In its own way, each discipline contributes to the creation of a *language game*, where although many of the words used sound identical to those that we use in common language, they acquire a different meaning which is dictated by that particular context. The *language game* is a concept coined by the philosopher Ludwig Wittgenstein to account for the fact that a certain word acquires a meaning dictated by the context in which it is uttered, and by the implicit or explicit rules shared by speakers in that particular context. Within a certain *language game*, a word acquires a meaning; in another *language game*, it assumes a different one. Let us take as an example the word "bill". If, while sitting in a restaurant with a friend, he says "I will pay the bill", it is safe to assume that he is referring to the fact that he will pay for the meal that you have just eaten. Whereas, if, during a parliamentary session, the Speaker of the House declares "The House will now vote on the bill", the word "bill" assumes the meaning of a proposed piece of legislation, and has nothing to do with the cost of a meal in a restaurant.

In the scientific domain, the meanings of words borrowed from ordinary language depend on that particular context. For example, the word "integral" in mathematics has a completely different meaning to the one it has in ordinary language. Additionally, science uses language creatively and invents new words, new concepts, or mathematical objects in order to account for particular phenomena and ideas. The word *soliton*, for instance, does not have any meaning in common language; it has been invented by scientists to classify and denominate certain peculiar wave phenomena (we will meet *solitons* later in our journey). This active and pragmatic operation of redefinition and the creation of new words allows light to be shed on parts of human experience that would otherwise remain hidden. This potentiality could be summarised by recalling the epigraph of this chapter, and by citing once more the Austrian philosopher: "If we spoke a different language, we would perceive a somewhat different world."

The words "stability" and "instability" in scientific jargon possess meanings different from those that they have in the *language games* associated with everyday life. However, despite their vagueness, the intuitions we obtain from common language do reflect some of the characteristics which are associated with the words in the context of mathematical sciences. For instance, let us imagine that in front of us there is a bowl with a marble sitting in its bottom. If the marble is pushed slightly in one direction with a tap of the finger, it will roll around in the bowl. It will oscillate a couple of times back and forth, and left and right, as it passes over the bottom of the bowl. After a while it will slow down due to friction. The marble will eventually lie still on the bottom of the bowl, exactly as it did before we disturbed it. This is an example of a system that finds itself in a state of stable equilibrium. Even if it is disturbed by an external influence, over the course of time the amount of disturbance will reduce, the perturbation will be dampened, and eventually the system will return to its original state. Now, let us consider what would happen if we turned the bowl upside down and placed our marble on top of it. If we give the marble a small nudge its resting state will be disturbed, and it will roll rapidly down the bowl's side. Its motion will cease far from its starting point. In this case we say that the marble at rest on top of the upturned bowl is in a state of unstable equilibrium. In a state of unstable equilibrium, the perturbations caused by a small disturbance are not damped over time, they grow substantially, and permanently move the system away from the state in which it was before the perturbation was introduced.

It is also possible to imagine a situation where someone swipes the bowl completely, taking the marble with it. This dramatic intervention certainly modifies the system through a powerful external influence. However in this book we will focus on instabilities of equilibrium states subject to small perturbations, which are exponentially amplified resulting in dramatic changes to the system's behaviour. Starting from simple intuitions like those we have just sketched, we will move towards the understanding of how the phenomenon of instability is conceived and studied scientifically, especially in physics, and what its essential features are. We will see how instabilities manifest themselves in many radically differing contexts, and how they give rise to a range of extremely rich and varied behaviours, far more interesting than the simple example of the marble and the bowl. Our first step in this direction takes us back in time by a few centuries, where we face a problem that has always fascinated humanity: the description of celestial phenomena.

CHAPTER 3

Planetary Instabilities

Raphael

In ancient rivalry with fellow spheres
the Sun still sings its glorious song,
and it completes with tread of thunder
the journey it has been assigned.
Angels gain comfort from the sight,
though none can fully grasp its meaning;
all that was wrought, too great for comprehension,
Still has the splendor of its primal day.

Gabriel

The earth as well revolves in splendor
with speed beyond all comprehending;
brightness like that of paradise
alternates with deep and awesome night;
the sea's vast floods surge up and break

DOI: 10.1201/9781003462040-3

in foam against the rocks' deep base,
and rock and sea are hurled along
in the eternal motion of the spheres.

JOHANN WOLFGANG VON GOETHE, FAUST I

So reads the epigraph of the majestic Prologue in Heaven of Goethe's *Faust*. In this series of vertiginous images, which emerged directly from the heart of German literature, the vision of planetary and cosmic phenomena is conveyed as consisting of a perfect movement, eternal in time, originally moulded by God himself. There is no trace of any instability whatsoever. Here we are led to the conclusion that we are in the presence of an eternal order characterised by the regular movement of the stars. Nineteenth and twentieth century physics has shown us something very different. We have discovered that the solar system is not stable, that the orbits of planets are not eternal, we have realised that their motion will one day become so chaotic that they break from their orderly procession through the heavens and might come crashing together! As we will see later, this catastrophe is, fortunately, not something we need worry about today.

We will eventually get to the details of this dramatic planetary instability, but before we do, I would like to guide you along a path through history. I want to show how we have moved from a view where all celestial phenomena were considered to be of a qualitatively different nature to terrestrial ones, to a conception where the same laws hold in heaven as on earth and how this has revealed the precariousness of planetary orbits and of the solar system.

The view that considers celestial phenomena to be of different substance and dynamics to earthly ones is ancient. In Aristotelian cosmology, the immutable and perfect heavens are counterposed to the Earth, the realm of change and imperfection. The Greek philosopher Aristotle had divided the world in two. The sublunar region, Earth, is in the processes of change, of generation and of death, due to the combination of four elements: earth, water, air, and fire. Conversely, the celestial world is constituted by a different element called ether. Ether, for Aristotle, is perfect, non-perishable, and immutable. Because the celestial realm is perfect, the bodies in it must move with uniform circular motion. In Aristotelian cosmology, the Earth occupies the centre of the solar system and of the Universe itself. The Sun and the other planets are fixed on concentric spheres rotating around the Earth. The Universe is finite and limited by the most distant sphere, where fixed stars are nestled. The Ptolemaic system – so called in honour of the astronomer Claudius Ptolemaeus who lived in the second century A.D. – integrated Aristotelian cosmology, which advocated for a geocentric and anthropocentric theory of universe, in order to explain certain phenomena which apparently eluded such a cosmology, for instance the retrograde motion of planets and apparent variations in their dimensions. Furthermore, the Ptolemaic system provided a practical instrument for the calculation and prediction of the motion and position of planets in order to create calendars or devices to facilitate navigation. One of the key concepts used in that system is the epicycle, a small circle along whose circumference a given planet moves. The epicycle moves in turn along the circumference of a circle with greater radius called a deferent. The Ptolemaic system is rich with such *ad hoc* hypotheses, which try to account for observable phenomena while preserving the Aristotelian view of the cosmos.

A very different model was proposed by Nicolaus Copernicus in the book published just before his death in 1543: *De revolutionibus orbium coelestium* (*On the Revolutions of the Heavenly Spheres*). Copernicus did not consider the Earth to be at the centre of the solar system. For him, the Sun was immobile and located close to the centre of the solar system. Its position corresponded to the centre of Earth's orbit, and around it the Earth and the other planets followed circular orbits. This was a radical step away from both the model of celestial motion accepted by physics at the time and from the Ptolemaic view. Between 1609 and 1612, the German astronomer, astrologist, musical theorist, and mathematician Johannes Kepler published three fundamental books: *Astronomia Nova* (*New Astronomy*), *Harmonices Mundi* (*The Harmonies of the World*), and *Epitome Astronomiae Copernicanae* (*Epitome of Copernican astronomy*). In these works, Kepler presented the three natural laws that bear his name. By carefully analysing precise observations of the planets' motions that had been reported by another scholar, Tycho Brahe, Kepler reached the conclusion that the orbits of the planets were not circular but elliptic in shape, and that the Sun occupied a locus called the focal point within these ellipses. In this way he was able to overcome many of the difficulties associated with the Copernican and of the Ptolemaic systems, which considered the planets' orbits to be perfectly circular. Kepler speculated that it was the Sun that caused the planets' motion through the emission of a peculiar substance and that, in turn, planets emit a different substance which attracts them to the Sun.[1] But, what is the explanation for the elliptical form of Kepler's orbits?

Kepler's observation found a dynamical justification based on the physical causes of the motion only later, after Sir Isaac Newton – mathematician, philosopher, alchemist, President of the Royal Society, as well as Master of the Royal Mint – had formulated his law of universal gravitation, which was published in 1687. This law states that two massive bodies attract each other with a force proportional to the product between the two masses and inversely proportional to the square of their distance. In the same treatise *Philosophiae Naturalis Principia Mathematica* (*The Mathematical Principles of Natural Philosophy*), Newton also formulated the so-called three laws of motion, on which classical mechanics is grounded. The second of these laws is the famous equation $F=ma$, which tells us what the force F will be if applied to a body with mass m so that its state of motion changes with acceleration *a*. Using the law of universal gravitation and Newton's second law of motion, it is possible to mathematically calculate the path of the orbit of the Earth around the Sun and to prove that the trajectory has an elliptical shape. This step, in which the forces at play in the attraction between the bodies leads to the trajectory, was accomplished for the first time by the great mathematician Johann Bernoulli in 1710.

In this way Kepler's laws acquire a foundation in the laws of dynamics. Furthermore, they show that the law that is used to calculate and explain the motion of massive bodies on Earth also applies to celestial bodies. Essentially, celestial and terrestrial phenomena were unified under a single description. This unification was accomplished by a movement that had started with the telescopic observations of Galileo Galilei more than a century before. Indeed, Galileo had observed that the Moon was not perfectly spherical or lacking irregularities in conformity to divine perfection; it has mountains and "seas" and its surface is corrugated just like the earthly one. He had also observed the presence of the so-called sunspots on the Sun which showed that, even there, there was activity and change. On top of this, he had verified the existence of the phases of Venus – which proved that it orbited around the Sun – the presence of Jupiter's satellites, and lastly of remote stars, the stars constituting the Milky Way, which had previously been presumed to be fixed and part of our solar system. Galileo had presented evidence in favour of the idea that the celestial world was similar in nature to the terrestrial one, and that it too was subject to change. Galileo and Newton had turned the ancient view upside down by showing, firstly, that change and "imperfection" involved not only terrestrial phenomena but celestial ones too, and secondly that the laws of change were the same in the heavens as on Earth.

Before we get to planetary instabilities, it might be a good idea to say a few words about the implications of Newton's laws of mechanics on the scientific view of the world at the time, as they will prove to be useful in better understanding how the conclusion that our solar system is unstable was reached. The powerful development of rational mechanics, by giants of the thought like Jean-Baptiste le Rond d'Alembert, William Rowan Hamilton, Leonhard Euler, Pierre-Simon Laplace, and Joseph-Louis Lagrange, enabled Newtonian mechanics to be integrated into a more general and powerful mathematical system. It also instilled into the minds of natural philosophers the *hubris* that they could predict the future of every single physical entity in the universe through the resolution of the respective equations of motion. On paper, the plan was very simple and can be summarised, roughly, like this: 1. Formulate the equations of the motion of massive bodies based on the forces at play; 2. Know the initial conditions (position and velocity) of the bodies at the instant of time $t=t_0$, from which one wants to calculate their motion; 3. Solve the equations of motion whose unknowns would be the positions of each body at every instant of future time. Indeed, given initial positions and velocities of all material bodies – provided that they can be known, and assuming that an excellent observer could record them all – the positions and velocities of the bodies at issue would be univocally determined for every instant of time t subsequent to t_0.

Stated like this, this all seems theoretically possible, if a little tedious. We have a list of instructions which employs the carefully collected data of the present in the meticulous and exact prediction of the future. According to this powerful view, the future is nothing but a trivial corollary of the present, nature is a great machine whose gears move according to Newton's laws. Obviously, the proponents of this conception did not expect it to be a simple task. To know the positions and velocities of all massive bodies in the universe at a particular instant in time was a practically impossible task, but what counted was the spirit of the enterprise, the idea that, if someone had knowledge of all the initial conditions, then thanks to the powerful edifice of analytical mechanics, they could know in principle the future of the universe in the smallest detail. Classical analytical mechanics is prescriptive, for it, the future is given once for all; it is closed. Unfortunately, or fortunately, depending on your viewpoint, this dream would never come true. Although writing the equations of motion for a particular physical system is possible, the knowledge of the initial conditions – the positions and velocities of every part of the system at a given instant in time – and most importantly the exact solutions of the equations is prohibitive and, in most cases, practically impossible.

Shortly, we will arrive at the failure of the great project of mechanistic determinism and its view that there is a possibility that we can predict the future with complete precision. But first, this may be a good time to introduce instability, as it will be our constant companion on this journey. In parallel to the developments of Newtonian mechanics and its application to the study of planetary motion, an interest in the understanding of celestial bodies and the stability of their orbits had begun to make its way into the scientific community. It was James Clerk Maxwell, the author of the celebrated – at least among physicists – *Maxwell's equations*, which describe light as an electromagnetic wave, who in 1855 performed the first stability study of a celestial system. At the time, Saturn's rings had been observed by telescope, but their exact nature was not known. The University of Cambridge announced the Adams Prize, named in honour of the man who discovered the planet Neptune, which invited candidates to consider three distinct possibilities: that the rings were rigid; that they were fluid or in part gaseous; or that they were composed of different isolated masses. The candidates were also asked to mathematically calculate in which cases they would be stable. Maxwell showed that only the third hypothesis allows a possible stability. On the basis of a mechanical calculation, he had theoretically anticipated the observational discovery that Saturn's rings are, indeed, constituted by numerous distinct masses orbiting the planet. To obtain his result Maxwell applied a mathematical procedure called stability analysis for the first time. That is, seeing that the equations describing the motion of the rings were far too difficult to be solved exactly in a general fashion, he considered *in primis* a simple solution corresponding to the regular motion of the rings, and to this solution he added further terms describing the effect of perturbations or deviations from that motion. In doing so, he managed to obtain simpler and approximated equations which still contained all the information about the systems' stability and still described the evolution of its perturbations, but which could be solved. In this way Maxwell was able to say in which cases the perturbations amplitude would grow enough to substantially change the motion of the rings and to produce instability, and in which cases the perturbations amplitude would decrease and eventually leave the rings' motion stable. Maxwell's hypothesis about the composition of Saturn's rings, based on the stability analysis, was later confirmed by astronomical observations.

If this was where the story of instability in planetary systems came to an end, it might seem like we were making a mountain out of molehill, but, in reality, this is just the prelude. To get into the thick of things we have to come back to the determinism problem which we hinted at previously, and to the possibility of predicting with absolute precision the future of a mechanical system given its initial conditions and the equations describing its motion. Towards the end of nineteenth century a *coup de grace* to the certainty of future prediction through classical mechanics was delivered by Henri Poincaré, a genial Frenchman, mathematician, physicist, engineer, and philosopher of science. He studied an apparently simple problem that had tested, and was still testing, several of the best mathematicians and astronomers: the motion of three massive celestial bodies interacting via the gravitational force. These might be, for example, the Sun, the Earth, and the Moon. Poincaré asked himself whether it was possible to find the exact solution to Newton's equations for the three bodies, a solution that would provide information about their orbits, namely the position of each of them at every instant in time. At the risk of curtailing your confidence in the path we have chosen to take, let it be said that the three-body problem can be solved exactly only in certain, very particular cases, and that even in those cases the solutions consist of mathematical expressions which have an infinite number of terms. As you can imagine, this constitutes a limitation for their practical use.

However, through this conceptual effort, Poincaré had revealed that a strong dependence on initial conditions could exist in the planets' motions. He had discovered, give or take, the following: If the three celestial bodies began their motion from certain positions and with certain initial velocities, then their orbits would have approximately a given shape. If the initial conditions were even slightly different, by just an infinitesimally small quantity, the resulting orbits would be completely different. Very tiny differences in the initial conditions correspond to huge differences in the dynamics of the system. Indeed, it is always possible to specify a motion's initial conditions with arbitrary numerical precision, for instance by saying that the value of the motion starting coordinate of a body with respect to a given reference systems is $x=2$ meters, or $x=2.00001$ meters or $x=2.000009321$ meters. This implies that in a mathematical calculation it is possible to obtain different orbits as the initial conditions are more and more precisely specified. Even from an experimental point of view, limits on prediction capability are encountered; all instruments used for measurement, including the most sophisticated ones, have finite precision. This implies that even by measuring the initial conditions of a particular system in a very precise way, one could not know them with arbitrary precision. This undismissible ambiguity leaves the way open for multiple distinct results for different versions of what appear, on the face of it, to be the same experiment. As, in a system like the three-body one, minuscule differences give rise to orbits which are radically different to each other, the system's dependence on initial conditions makes it impossible to predict the resulting motion in the long term. As we will see in more detail later, dependence on initial conditions is one of the salient features of chaotic systems.

Poincaré had opened the way to the study of chaos theory that was to develop during the course of twentieth century.

From a practical point of view, if one is happy with a solution to the equations having limited validity, it is still possible to obtain approximated shapes of a planet's orbit around the Sun, taking into account the effect of other planets or of comets and other celestial bodies too, through a method called perturbation theory. However, this information is not exact and, despite being sufficiently precise in the short term, can not provide any absolute guarantee about the distant future. Epistemologically, Poincaré's discovery had revolutionary repercussions and is one of the blows that lead to the cracking of the deterministic paradigm or, better said, to the decoupling of the concept of the predictability of a system's state from the determinism of its dynamics as dictated by the equations of classical mechanics.

And here we are close to grand finale. Now we are in a position where we can ask ourselves, given the fact that we do not know exactly how to solve the equations of motion for the Sun, Earth, and Moon, and even less so the ones for the motion of the entire solar system, should we perhaps renounce any claim that we know whether our solar system is stable or not? Shall we never know if the planets will continue to travel their orbits or if they will deviate from them?

Not exactly. The advent of computers has been of help. In the final decades of the twentieth century, it became possible to perform massive numerical experiments which allowed electronic computers to solve the equations which describe the evolution of the solar system. The most recent results are astonishing. The solutions found by computers show that the solar system is unstable. According to this research, the instability will manifest in dramatic changes including the possibility of collisions between planets, and planets being ejected from the solar system itself. These catastrophes should occur in times of the order of several millions of years from now. Despite these dates seeming very far from the present, what matters, at least for scientists, is that such instability should occur within times inferior to the expected lifetime of the Sun, which is estimated to be approximately equal to five billion years, and the fact that the solar system's instability *could* cause it to behave in a chaotic way at all.

What a radical change of perspective from the ancient conception with which we began, according to which celestial bodies were of a different nature to earthly ones and were perfect and immutable! Copernicus, Galileo, and Kepler conferred movement and imperfection to celestial spheres, and, later, Newton discovered that it is the same laws which govern the fall of objects on Earth that govern the planets' orbits around the Sun. We have, then, arrived at the hypothesis of instability according to Maxwell, and of the chaos and ultimate unpredictability of the solar system introduced by Poincaré and developed further by current research: *sic transit gloria mundi,* sorry *astrorum*!

NOTE

1. It is interesting to recall that the idea of the Earth orbiting around the Sun was not completely new, but it had been in reality suggested by Aristarchus of Samos, who lived between the third and fourth century B.C.. Aristarchus held to be true that the Earth orbited around the Sun with circular motion, while the Sun and the stars were immobile.

CHAPTER 4

Cybernetics and Control Systems

Trust is good, control is better.

VLADIMIR ILYICH ULYANOV ALIAS LENIN

When, getting home from a walk on a chilly day, we turn the knob on the thermostat or press some buttons to start the heating, our first thought is not usually to marvel at how we are engaging with a system operating thanks to a mechanism of circular causality perfectly exemplifying some of the key traits of a scientific discipline called cybernetics. I also doubt that cybernetics crosses your mind when you drive your car or catch the train.

What, you may ask yourself, has this strange preamble to do with instabilities? In this chapter I would like to guide you through the discovery of practical instability. This is a type of instability that relates to engineering, and is much closer to home than the instability of the solar system in the distant future. The instability that I want us to investigate is the one of the control systems which underpin the operation of the myriad of machines which litter, and are interwoven with, our daily existence. As we will see, interest in this kind of instability began to flourish at

DOI: 10.1201/9781003462040-4

the same time that the first studies of the instabilities of the solar system were being made, and the two share some common protagonists. But I would like to approach this story starting from certain ideas about circularity which form the basis of past and present technology, and throw an *en passant* gaze on the famous cybernetics. Along this path we will collide with a paradoxical idea, according to which in order to fully understand and optimise the stability and control of technological and natural systems, we must make use of substantial approximation and simplified forms of our most general theories. We must take a practical approach which, by focussing on the particular, loses universality, rather than seeking to employ knowledge which speaks in a more universal fashion but which remains silent in front of peculiarity. Without the approximation process, the understanding of instability and the possibility of controlling it will forever elude us.

But, for the moment, let us return to that chilly winter afternoon. The thermostat, once set at a certain temperature, orders the heating to start warming up the house. The thermostat is connected to a thermometer, from which it receives a signal containing information about the temperature in the house. It compares that signal with the ideal temperature value set by the user. If the temperature in the house is too low the system will keep the heating on or even increase output, and if the temperature is higher than desired it will reduce the power or it turn it off. The behaviour that we have just described is an example of a circular causality process; a certain element – the thermostat in this case – causes a modification to another one – the heating system in this case – which changes the conditions of a third one – the environment, the house – modifying its behaviour. This second element then causes a modification to the first element, in turn, changing the first's behaviour. This circular dynamic is called a *feedback loop*. An unpleasant example of a *feedback loop* is the Larsen effect, which is well known to musicians. In this case, the sound emitted by an amplified speaker on stage is picked up by a microphone, which broadcasts it back to the speaker which amplifies it again, and so on in closed circuit. The result is an annoying whistle, whose frequency corresponds to the resonance frequency that is amplified the most in that circuit. However, not all *feedback loops* are as irritating as this one, and we can find many more which are extremely useful. The capability of a system to self-regulate is of vital importance when one wants a machine or a device immersed in a certain environment to maintain a given behaviour even if the environment influences the machine itself. This is even more important where the machine's behaviour transforms the external environment and this environmental transformation then affects the machine. For this reason, the concept of a *feedback loop* is at the heart of cybernetics, the discipline that was born at the end of the 1940s, and develops at the interface between dynamical systems theory,[1] anthropology, biology, engineering, and human sciences. The aim of cybernetics is to understand the mechanisms of control, communication, and self-regulation that exist within organisms – both living and mechanical – and of complex systems constituted by the interactions of many different parts. The term cybernetics was coined by the American mathematician Norbert Wiener, and it is etymologically connected to the Greek words κυβερνητική (*kibernetiké*) and κυβερνήτης (*kibernétes*), meaning control and helmsman – or ship's skipper – respectively.

In reality, *feedback loops* are present in almost all human activities connected to learning, and where a behavioural scheme of the following kind is needed: the execution of an action with the aim of reaching a given objective; observation of that action's result and its comparison with the objective; a modification of the initial behaviour following from the message that informs us about the result of the action; the execution of a new action informed by that message; and repetition of the above steps.

The fundamentals of this structure are also rooted in spiritual practices which predate the development of machines and sophisticated technology. For instance, it is an important part of Zen, a school of Buddhism introduced in Japan in the seventh century A. D., which flourished in China under the name of Chan having arrived there from India. In the meditation practice of *zazen*, or sitting meditation, the practitioner must sit immobile on the meditation cushion doing nothing in particular except maintaining a simple posture: the legs crossed in the lotus position with the knees touching the ground, the back straight with the spine forming a slight arc, the back of the neck pushing towards the sky, the chin slightly tucked, the mouth closed and the tongue resting on the palate, the eyes quiet and gazing towards the floor a few centimetres distant or pointed towards the wall in front, the shoulders relaxed, and the hands in the lap forming a *mudra*, where the thumbs are touching as if holding an egg. The mind should be empty of any thought, not drowsy but awake and in conscious observation of all the posture's salient points. When a discrepancy is noticed with respect to one of the posture's key points, an immediate action brings the posture back towards its ideal form. The mind, without judging, observes the result obtained. If another point of the posture has changed, it acts to bring it back towards the ideal form, then observes the result, it acts again, and so on. Absolute immobility consists of an endless cycle of infinitesimal adjustments. Conscious attention notices a posture point out of place and urges to action to bring it back into form, and so on to infinity. In this meditation a continuous *feedback loop* manifests; it is made of action, observation of the result, comparison of this result with ideal posture, and a new action.

But now we will return to the more prosaic issue of nineteenth century mechanics which, on the one hand, extended its gaze upwards to grasp and minutely predict the motion of celestial bodies, while with the other, it stretched down to harness the forces unleashed by the combustion processes of steam engines which, channelled through gears and metallic mechanisms, became motion. The introduction of steam machines during the Industrial Revolution made it necessary to develop self-regulation mechanisms that would ensure the reliable operation of engines. In this context it was necessary to know under which conditions these control mechanisms would ensure that the system remained stable. Namely, if and when the self-regulatory activity of control devices would bring the results that they had been designed for, and what could be done to prevent their actions from giving rise to behaviours that drifted from the operation goal.

It was our old friend James Clerk Maxwell, who, not satisfied with solving a stability problem in a system made of orbiting celestial bodies, successfully formulated the first mathematical study of the stability of a control system. In 1868 Maxwell sent a paper called *On Governors* to the Royal Society. At time of the Industrial Revolution it became necessary to have a reliable method of controlling the speed of new steam engines. Governors were the mechanisms designed for this purpose, and within them there was a *feedback loop*. A certain amount of steam – in this case the result of the action of the mechanism acting as control – was injected into the engine – the device to be controlled – in such a way as to increase its rotation speed. If undesired perturbations or variations in the power supply caused an increase of the engine speed to exceed a certain value, a mechanism would be activated with the goal of reducing the steam flux in the engine and thus its velocity. On the other hand, if the engine speed was lower than a desired value, the mechanism increased the steam flux entering the engine in order to increase its velocity. But how were these governors made? A typical example is the one designed by James Watt, the inventor of the steam engine itself. His governor consisted of a rotating axle, connected to the engine and activated by its rotation, to which two metallic spheres were attached by pivots. During rotation, centrifugal force meant that the distance of the spheres from the axle varied proportionally to the speed of its rotation. When the engine rotated too quickly and the spheres moved away from the axle, a mechanism connected to the sphere's pivots was activated.

This mechanism would slightly close the valve through which steam entered the engine. This reduced the rotation speed, which caused the rotating spheres to move closer to the axle again. If the engine speed diminished too much, the rotating spheres found themselves very close to the axle and the mechanism connected to them opened the valve to let more steam enter the engine. The governor was stable when environmental or internal perturbations did not substantially impede the engine's operation.

In Maxwell's paper we read these words:

> It will be seen that the motion of a machine with its governor consists in general of a uniform motion, combined with a disturbance that may be expressed as the sum of several component motions. These components may be of four different kinds: the disturbance may 1. continually increase, 2. it may continually diminish, 3. it may be an oscillation of continuously increasing amplitude, and 4. it may be an oscillation of continually decreasing amplitude. The first and third cases are evidently inconsistent with the stability of the motion; and the second and fourth alone are admissible in a good governor.

Essentially, for a governor to be useful and functional, it needs to ensure that possible perturbations are continuously dampened, and that this can happen even when some oscillations are present, provided that their amplitude reduces more and more in the course of time. In the following parts of this milestone of control system engineering, the Scottish physicist mathematically analysed, and in many cases identified, the stability conditions of various governors created by important inventors of the time like Carl Wilhelm Siemens, Fleeming Jenkin, William Thomson (Lord Kelvin), and Léon Foucault. The theory of dynamical system stability would later be formalised in a rigorous and systematic fashion by the Russian mathematician Aleksandr Mikhailovich Lyapunov, who lived between the end of the nineteenth and the beginning of twentieth century. It was developed further by the mathematician René Thom, with its catastrophe theory, and by other scientists including Edward Routh, Adolf Hurwitz, Hendrik Bode, Harry Nyquist, and Norbert Wiener, with particular reference to control systems and cybernetics.

The issue of stability of governors was relevant to the Industrial Revolution for purely practical reasons. Nowadays, the use of feedback mechanisms is decisive in engineering for the creation of effective machines and technological devices in the most diverse fields, from mechanics to aerospace engineering, from electronics to photonics. After all, what matters the most to us is not the particular governors problem tackled by Maxwell, but the idea that he used to analyse their stability. Here we reconnect with the discourse on stability analysis outlined in the previous chapter and explore it in a deeper fashion. By examining governors, Maxwell found himself analysing systems whose equations of motion, which describe its dynamics at any instant of time, are nonlinear. The word nonlinear has here a very precise meaning that can be expressed in a completely satisfactory fashion only by resorting to mathematical language. However, we can say that a nonlinear system is one whose response is not directly proportional to the stimulus it is subject to. If a nonlinear system is allowed to oscillate at a certain frequency, it may start to generate oscillations at other frequencies. A nonlinear system can be one which suffers an effect from the external environment depending on how the system itself had influenced the external environment in the past. Mathematical equations describing this kind of behaviour are said to be nonlinear. These equations frequently cannot be solved mathematically in an exact fashion and in general form. The intuition that by starting from the general equations of motion of the system it was possible to calculate further approximated equations describing the motion not of the complete system but simply of small perturbations added to the ideal behaviour of the system itself had already been used by Maxwell in his study of the stability of Saturn's rings. These new equations turned out to be linear; they were simpler and in general solvable.[2] For this reason, this technique is called the *linear stability analysis*. The brilliance of Maxwell's idea lies in the fact that all information relevant to the system's stability is contained in these approximate but solvable equations. His approximation was radical but not destructive: it highlighted the salient features of the stability of the system he studied. To this day, stability analysis of dynamical systems still almost always uses approaches similar to those used by Maxwell, deriving simple and approximated equations which describe only the perturbations' motion.

In this sense all the power and nobility of the concept of approximation in physics, and in sciences in general, emerges. In everyday language, the word "approximation" often has negative connotations. Approximation is associated with imperfection, incompetence, and sloppiness. In reality, most scientific theories – if not all of them – exhibit various approximations, frequently very drastic ones, but their accuracy remains well within narrow and well-defined margins. This is a fact that has possibly escaped the attention of the general public. People tend to treat scientific truth as if it were a sort of metaphysical truth, carved in stone. This is in part due to the fact that scientists are human beings who, under certain circumstances like the clerics and oracles who preceded them, are not always inclined to reveal all their secrets, and certainly do not wish to emphasize the weaknesses or the vulnerabilities in their theories when they present them to the general public or when they wish to obtain funding for their research.

"*La física es el arte de la aproximación*" – "Physics is the art of approximation" – I was frequently told in classes given by Germán de Valcárcel, professor of quantum and nonlinear optics at the Universitat de València, and in-between long discussions with my eyes staring at the blackboard trying to disentangle the mathematical intricacies of the symbols describing the operation of a self-pulsating laser. I was in the last year of my Master's in Physics, almost ten years ago, and I spent it with his research group. Maxwell's example is without any doubt a plastic embodiment of this concept. There, approximation simplifies a problem's solution, making it accessible while at the same time preserving, within the simplicity, its essence; approximation allows its understanding. The use of approximation is both a pragmatic solution to the problem and an inspired intellectual leap. It is an act of courage, signalling a willingness to lose the generic view of the whole in order to focus on a part of it. It is a movement that trims down the superfluous and eventually highlights the most intimate nature of the phenomena.

This resorting to approximation, this having to reduce to essential terms and losing information about the more general context in order to obtain a particular result, is not confined to knowledge processes. It is also necessary in many of our practical and daily activities. To act in an efficient way, we frequently have to forget, to sacrifice, the whole. We have, instead, to focus on the particular, operating with it, dancing with it. This step, where we lose sight of the totality and lose interest in its destiny or in the consequences that our loss of consideration could entail, is frequently difficult to accomplish. The dilemma of what we should do, the uncertainty of the moment preceding the action, asking oneself whether the damage that will arise from what one loses as a result of the action will outweigh the benefits, the fear of the fact that action unavoidably disrupts the harmony of the whole and brings with it error and suffering, sends my thoughts back in time to epochs past. In the *Bhagavad-Gita*, the heart of the Indian epic poem *Mahabharata* and a sacred text of Hinduism, Krishna says to Arjuna, who fearfully asks about the consequences of his action before taking the battlefield: "Therefore, always perform unattached the deed to be done, for the man performing action while being unattached attains the Supreme." On the contrary, in the case of the scientific study of instability, one acts moved by interest – attachment – for the knowledge and control of a particular system, and the action which is necessary for the success is the one of approximation, which lets loose from sight the whole and what is most universal.

NOTES

1. We can say in a simplified way that a dynamical system is a physical, chemical, engineering, or biological system where the features describing it (for instance the position of a particle, its velocity, the density of a given substance, the intensity of an electromagnetic field, and so on) change in time according to rules expressed in form mathematical differential equations.

2. A linear relation between two quantities y and x can be expressed by the following function $y=ax+b$ where a and b are coefficients. Geometrically the function $y(x)$ is represented by a line on the Cartesian plane, hence the term "linear". A nonlinear relation between two quantities y and x could be, for instance, instead $y=cx^2$ or $y=dx^3$ where c and d are coefficients In general, the evolution equations for n coupled perturbations amplitudes a_i (with i an integer ranging from 1 to n) can be written as $\dot{a}_i=L_i(a_1, \ldots, a_n)+N_i(a_1, \ldots, a_n)$, where the · symbol denotes the first derivative with respect to an evolution coordinate – typically time or space – and L_i and N_i are linear and nonlinear functions of the various perturbation amplitudes respectively. When N_i is neglected for all i, the equations for the perturbations become linear and are said to have been "linearised".

CHAPTER 5

Hydrodynamic Instabilities

Chaos and Turbulence

I observe the motion of water surface, which resembles the movement of a head of hair, which on the one side is due to the hair weight, on the other by the locks shape. So the water follows a curvy motion, which partly follows the direction of the current, and partly a random and opposite motion.

The water which from the mountains peaks till their roots descends, at any degree of its descend it acquires a degree of turbulence.

LEONARDO DA VINCI

Leonardo da Vinci was probably one of the first people to study the phenomenon of turbulence, not only reporting his findings in words but also meticulously reproducing, in a series of amazing drawings, the richness of the dynamics of water's swirling and curving movements. Leonardo lived from 1452 to 1519, and

DOI: 10.1201/9781003462040-5

in those few years he made extraordinary contributions not only to the art of painting, but also to engineering, hydrodynamics, anatomy, and zoology. He was also a musical virtuoso. In this chapter I want to talk about how turbulence as a process is connected to instability phenomena, and how instabilities manifest themselves diffusely in fluid dynamics by giving rise to surprising coordinated and chaotic behaviours.

The study of turbulence resumed some centuries after the works of Leonardo, in the nineteenth century to be precise, thanks to fundamental contributions by the mathematician and physicist George Gabriel Stokes and by the mathematician and engineer Osborne Reynolds. Reynolds observed experimentally that a fluid flowing through a pipe exhibits different behaviours depending on the velocity at which it travels. At low velocity the fluid can be described as a series of thin layers which flow parallel to each other moving in the same direction, without crossing or mixing. In this regime, called laminar, the fluid remains orderly and steady; it is the viscous friction forces among its component particles which dominate the dynamics. When the fluid velocity exceeds a certain critical value, the orderly and parallel flow of these layers changes suddenly and the fluid becomes unstable, its appearance mutates radically, and vortexes and eddies arise. This is the turbulent regime.

Turbulence is a complex process where chaos and order coexist. In an apparently erratic motion one can distinguish ordered forms which physicists call coherent structures. These take the form of rotating vortexes inside the global fluid flow. Once the instability of the laminar flow is activated, energy is transferred to smaller and smaller vortexes in a cascaded like process, until it is eventually dissipated as heat due to friction. This cascading process of energy transfer from large whirling structures to smaller and smaller ones was analytically characterised by the Russian mathematician Andrey Nikolaevich Kolmogorov. Kolmogorov mastered several different fields of scientific knowledge: establishing the axioms of probability theory, founding the study of computational complexity, and also making decisive contributions to classical mechanics. His main work in the field of turbulence consisted of identifying, in a quantitative and universally valid fashion, that the energy transfer occurs from large scale vortexes towards smaller scale ones, and in determining the smallest dimension of the vortexes forming in the turbulent motion as a function of the fluid parameters, namely the end of the energy cascade before it is dissipated as heat.

Turbulence manifests itself in many situations that will be familiar of every one of us. We see it in the fluxes of rivers and streams that Leonardo observed, but also in the motion of cigarette smoke, where we see an initially ordered and directional flow gradually spreading through a space in spirals of twisting, whirling motion. Elsewhere in daily life, turbulence can be seen in the emissions from cars' exhaust pipes and in the trails left by the air jets emitted by airplanes' turbines. The erratic beauty of turbulence's whirling can be beguiling, but understanding its formation is of vital importance in various engineering contexts where it needs to be avoided. However, fluids' instabilities do not only appear in turbulence processes but in many other situations too. The entire study of fluid dynamics is awash with a vast number of instabilities that it would be impossible to discuss here. We will limit ourselves to a fleeting glimpse of some paradigmatic examples as if they rush past us in the mountain stream.

Let us start in 1831, when Michael Faraday first described the process in which an homogeneous fluid layer in a container which is shaken vertically becomes unstable and develops, on its surface, waves whose frequency are half of the one at which the container moves up and down. These waves form oscillating spatial structures and were later observed in many systems subject to periodic change, for instance in particular granular media, in quantum gases, in chemical reactions, and even in optics.

You may think that the instabilities of fluids are only of academic interest, exciting for scientists locked up in their ivory towers or in their laboratories but of no interest in the real world, but you could not be more wrong. Cyclones and typhoons, giant vortexes which have radii of several tens, if not hundreds, of kilometres and are associated with adverse meteorological conditions, like extreme winds, tornados, and violent storms, form in the terrestrial atmosphere in a set of dynamics in which the *baroclinic* instability plays an important role. Staying on the theme of vortexes, it is important to mention the Kelvin–Helmholtz instability, which occurs when two fluids that are in contact with each other flow with different velocities. This results in the amplification of perturbations at their interface, giving rise to rotating waves which eventually mutate into a whirling motion. The waves generated can have macroscopic dimension and can be seen in the cloud structures that form in the terrestrial atmosphere and also in those of other planets in our solar system.

Where, instead, one has two fluids with different densities in reciprocal contact, the Rayleigh–Taylor instability can arise. In this case, if, for instance, the heavier fluid is located on top of the lighter one, then the state in which the fluids are separated becomes unstable and a mixing process takes place, which is followed by a whirling dynamic. A spectacular and terrible example of this instability is the creation of the so-called mushroom clouds of smoke and debris that follow the detonation of nuclear warheads. The nuclear explosion creates a burning sphere, the interior of which has a temperature that can reach several hundreds of millions degrees Celsius. These temperatures are higher than those generated at the centre of the Sun. The burning sphere ascends, creating a void. As it expands, hot air is sucked into this void's interior, which then in turn expands. As a result of heat, the air inside the sphere becomes less dense than the air that surrounds it; this is necessary for the Rayleigh–Taylor instability to manifest. A spontaneous rotation process is triggered. The process of ascension flattens the top of the sphere, which then falls back on itself due to its interaction with the surrounding air, where the density difference between interior and exterior is no longer substantial. Elsewhere, the Rayleigh–Taylor instability can be found in the dynamics of stellar explosions, like those of supernovae. Here vortex dynamics acquire a planetary scale. The effect also manifests in technological contexts, where the force of nature is harnessed with non-destructive goals, for instance in nuclear fusion reactors. The aim of these devices is to generate energy through the fusion process of hydrogen isotopes,[1] which leads to the formation of heavier atoms. This process frees an energy excess that can be used for practical purposes. Nuclear fusion is the engine driving our Sun and other stars. In order for this process to work in an optimal fashion, and for it to be sustainable, it is necessary to keep the reacting matter confined in a limited region of space, which occurs in stars easily because of their powerful gravitational fields. This is necessary as the fusion reaction probability is proportional to the density of the reacting atoms. Fusion is brought about by a temperature increase in the atoms which, at a certain point, become ionized and lose electrons, which are torn away from them due to the high level of excitation provided by thermal energy. In this way a plasma is formed: a gas of positively charged ionized atoms and electrons. From now on nuclei can fuse together, freeing energy. When this process occurs in a laboratory, the plasma which forms must remain confined in a certain region – or cell – of space, a situation realised using special laser light beams or through the application of intense magnetic fields. However, the Rayleigh–Taylor instability lies in wait and can amplify small undulations and corrugations on the fuel cell's surface, altering the shape and potentially causing an irremediable reduction of the fusion process' effectiveness. Therefore, this important phenomenon must be taken into consideration when designing efficient fusion reactors.

Another extremely relevant hydrodynamic instability is the Rayleigh–Bénard one. Let us consider a fluid located between two parallel surfaces. If it is kept at a constant temperature, the fluid finds itself in a state of stable equilibrium; any small fluctuations connected to possible macroscopic motion inside it are rapidly dampened and the *status quo* is preserved. However, if the temperature of the lower surface is raised, the fluid's rest state becomes unstable. As it is warmed by the lower surface, the density of the bottom most portion of the fluid is lowered and it tends to rise, while the fluid portion located close to upper surface, being cooler and thus having higher density, tends to fall. When gravitational forces overcome the resistance of viscous friction, an ordered and large scale circular movement establishes itself. What is called a *convective cell* is formed, a flux of rotating fluid beginning at the bottom, which on reaching the upper surface is cooled, and then descends once more. This rotation can occur in both clockwise and anti-clockwise directions, but where cells form close to one another, if one cell rotates clockwise its neighbour will rotate in the opposite direction and vice versa.

While the convection process caused by the Rayleigh–Bénard instability has a quasi intimistic dimension and the liquid warmed up from below can resemble the daily operation of boiling water to cook pasta, the same cannot be said of the convection process when it manifests itself in the terrestrial atmosphere. The American meteorologist Edward Lorenz, at the beginning of 1960s, studied this phenomenon through a mathematical model which bears his name to this day. He observed, through a computer simulation, that one solution of the mathematical model was unstable and that the resulting instability gave rise to an irregular and non-periodically repeating behaviour in the system. The system became unpredictable and erratic: it was chaotic. Although Lorenz's equations are deterministic, in that the value of physical variables at a certain instant of time is determined from their value at the preceding instant following fixed rules, the actual results show that it is impossible to predict the system's behaviour at a certain instant in the far future. The combination of the deterministic component of the equations and the chaotic behaviour led to the phenomenon observed by Lorenz being denominated *deterministic chaos*. We cannot give here a detailed technical definition of what is meant mathematically when one speaks of a chaotic systems; however, one of the salient features of chaotic systems associated with their unpredictability is the strong dependence of their behaviour on the initial conditions of their evolution (do you remember Poincaré?). If we imagine that we know exactly the initial conditions of the system, we can compute its state at a certain later instant far in the future. However, if we now repeat our calculation better specifying the initial condition, for instance by adding a further decimal number to the value used in previous calculation, we will find that we obtain a completely different result. We can repeat again and specify our initial condition arbitrarily until using the millionth decimal digit. Even after the last approximation, if we add a further decimal digit and we calculate the system evolution again, we could obtain a completely different result. Very tiny changes in the initial conditions give rise to distinctly different later states. This extreme sensitivity to initial conditions makes these kinds of systems actually unpredictable. The famous term *butterfly effect* was coined to describe just how the dependence of a process on its initial conditions could lead to completely different effects in the long term. It is a metaphor which describes the minuscule wing beat of a butterfly in a certain place as the potential cause of a tornado or of majestic and giant atmospheric phenomenon manifesting on the other side of the world.

Another important feature of chaotic systems is the existence of islands of predictability in their behaviour called attractors. Namely, the system's behaviour is unpredictable in the single case; however, generally, we can know roughly in which configuration all the possible states available for a given set of parameters the system will find itself. In the course of time the system "draws" an orbit, leaves a trace, in this mathematical space called *phase space*. In the case investigated by Lorenz, the trajectory does not cover it entirely but it stays only in certain well-defined areas. The ensemble of the *phase space* zones where the system lingers constitute what is called an *attractor*. This means that the system prefers a certain type of behaviour, certain values of the physical variables that characterise it. Lorenz's work constituted a milestone for the development of chaos theory. Deterministic chaos was later observed and studied in many different contexts like optics, chemical reactions, and electrical circuits. In Lorenz's model, instability constitutes the incipit of chaos, the spark setting ablaze the dynamics of unpredictability and unleashing the absence of certainty.

In this chapter we have briefly immersed ourselves in the whirling landscape of fluid dynamics' instabilities. We have travelled from the fifteenth century and Leonardo da Vinci to the atomic explosions of the twentieth century, sailing through incandescent plasmas, and ending with chaos in meteorology. We have seen how turbulence and chaos arise from instabilities of fluids' fluxes and how the seed of infinitesimally small perturbations can completely transform the character and the behaviour of unstable systems. But it is not over yet. We have not yet even hinted at instabilities of the phenomena which possibly come most easily to mind when one thinks of a fluid: waves. In the next chapter we will glimpse the instabilities of waves in a liquid medium and of those of light waves too, considering some examples of what is known as *modulation instability*.

NOTE

1. Isotopes are two or more atoms having the same number of protons, and hence belonging to the same chemical specie, but different regarding the number of neutrons in their nucleus.

CHAPTER 6

Modulation Instability

Theoretical Physicist

Like a minuscule grey-coloured cricket,
A wee man by a blackboard is clicking.
That's his way to invite for a mosey
Across worlds that not many find cosy.
But as soon as it comes to departure,
The attendance appears rather patchy:
Only few still persist at note-taking!
Tree of knowledge, thy fruit is bliss-making...
Keep your hands off our light entertainment,
Do not tempt us with crumbs of attainment,
Do not teach us the right aspirations,
Do not tease us with serving the nation.

VLADIMIR E. ZAKHAROV

DOI: 10.1201/9781003462040-6

Who knows if Nikolay Nikolayevich Bogolyubov regretted disregarding that minus sign? The scandal happened in an article entitled "On the theory of superfluidity", published in 1947 in the *Journal of Physics of the USSR*. In the article Bogolyubov mathematically studied the properties of superfluidity – the ability of a fluid to flow without dissipating energy due to friction – in a particular quantum gas called *Bose–Einstein condensate*. At temperatures close to absolute zero, the particles that constituent the gas "lose" their individuality and the gas becomes an entity called *condensate* that can be described as a single great wave capturing its collective dynamics.

The model used by Bogolyubov to understand the behaviour of Bose–Einstein condensate is a particular equation describing the reality of a variety of different physical systems called the nonlinear Schrödinger equation.[1] While studying the mathematical properties of that equation he noticed that one of its solutions, which describes a spatially homogeneous distribution of the gas, was unstable if a certain physical quantity had a negative sign. However, it was more important to focus on the case with a positive sign for the particular phenomenon of interest to Bogolyubov in the article. So, he simply noted that in the case of the negative sign there would be an instability of the solution describing the homogeneous distribution of the gas, and that the amplitude of some small oscillations would grow exponentially, and moved on to study the properties of the stable solution to the equation. If he had pondered longer over this minus sign he would have discovered, in detail, the properties of modulation instability. Or maybe he did discover them, but never published the results.

We can represent modulation instability in the following way: let us imagine a wave of constant amplitude that oscillates at a given frequency which propagates on the surface of sea, and let us imagine small wavelets overlapping it that have minuscule amplitudes oscillating at a different frequency. Modulation instability consists in the fact that the main wave, while propagating, cedes energy to the wavelets which grow exponentially in amplitude creating a modulation of the original wave until the latter is completely destroyed. The original wave is hence unstable with respect to the presence of these small wavelets which, acting initially as perturbations, grow in amplitude and subtract energy from it.[2]

Now, this instability manifests itself in the case of sea waves, but it is also present in other contexts that seem unconnected. One of these is the case of a powerful light wave propagating along an optical fibre. An optical fibre consists of a sort of thin pipe made of silica – a glassy material – along which light waves can flow. This peculiar pipe has a core with a very small diameter, generally of the order of a one hundredth of a millimetre. This is surrounded by a less dense glass that has a diameter of about one tenth of a millimetre. These are both enclosed in a protective jacket. An optical fibre can be anywhere from a few centimetres to hundreds of kilometres long. Light waves generated by a laser are injected at one end of the fibre and propagate along it inside the core until they exit at the other end. Incidentally, optical fibres constitute the skeleton of today's information society, as almost all the data traffic which enables the existence of the internet travels through them. If a light wave oscillating at a certain frequency begins its journey at one end of the fibre and we find that it is almost identical at the output end of the fibre, we say that the wave is stable during propagation. If the wave breaks up during its propagation and at the output end a distribution of light very different from the one at input appears, for instance, one characterised by an alternation, in time, of moments of light and darkness, we say that it is unstable. If the fibre and frequency of the powerful input wave are chosen in such a way to satisfy certain particular conditions, here the small light wavelets oscillating at different frequencies and fluctuating on top of the original wave gain exponentially energy from the latter upon propagation until they become greater than it and it is destroyed. These small wavelets, acting as destabilising perturbations, are caused by the optical noise of the laser source used or by quantum fluctuations, and they are essential to the destabilisation of the powerful initial wave.

Modulation instability understood in a general sense does not concern only waves; it is an instability where the homogeneous state of a system – constituted for instance by a constant amplitude wave or by a uniform distribution of matter in space – is destabilised and destroyed by oscillations modulating it through the creation of some non-homogeneities. These are areas where a certain physical quantity has a stronger presence and other areas where the physical quantity is scarcely present or even absent. Some examples of systems in an homogeneous state are a stretch of water where the fluid's surface does not exhibit oscillations, a beach where the sand is uniformly distributed, a light beam whose intensity is constant and uniform, a wave travelling in a fluid with a constant amplitude, or a chemical reaction where the reactants are uniformly distributed inside the container. Let us now imagine that one of these homogeneous states is perturbed, namely that for some reason small oscillations or non-homogeneities form in it. For instance wavelets on the stretch of water, a non-uniform sand distribution with dunes and ditches on the beach, areas where light is more intense and others where the beam is less concentrated, or zones where some of the chemical reactants are more concentrated and other less inside a container. We will say that the homogeneous state is stable whenever the modulations and non-homogeneities of these waves are not amplified over the course of time, but rather remain unchanged or are dampened and disappear. We will say that the state of the system is unstable when non-homogeneities are gradually amplified over the course of time until they become so significant that they radically transform the original landscape or cover the tracks of the steady primitive uniformity. In such cases we would have a water surface where the fluid level changes substantially in different regions of space, a light flux where areas of light and darkness clearly alternate, a sandy surface where dunes and ditches appear regularly spaced as if someone had drawn them dragging a stick through the sand, or a chemical reaction where the molecules of one of the reactants are all concentrated in one area of the container and the molecules of the other are concentrated in a different area, or a wave whose amplitude is not constant and uniform but exhibits areas of greater intensity and others of lower intensity.

However, modulation instability does not always give rise to reordered states. Examples of regular order, repeated *patterns* of the kind we have just described can be short-lived and later give way to highly chaotic and disordered states. What is fascinating about modulation instability is that in certain cases the system's homogeneous state vanishes and temporarily leaves room for modulations, to non-homogeneities, but after a while the system's homogeneity returns. Non-homogeneities are dampened and for some time the system returns to its original steady state. Then, non-homogeneities prevail again. This surprising alternation can repeat itself multiple times before the non-homogeneities eventually get the upper hand for good. This return of systems to homogeneous states is called *recurrence*.

Modulation instability, after its quasi-discovery by Bogolyubov, was rediscovered and studied specifically in the 1960s by Brooke Benjamin and Jim Feir in fluids, by Bespalov and Talanov in optics, and then by Vladimir Zakharov and his school of thought in different physical systems ranging from fluids to plasmas and to light waves propagating in solid material. The decisive contribution of the Japanese scientist Akira Hasegawa was his experimental observation in optical fibres and suggesting the possibility of using it as a method to amplify optical signals for practical purposes. We will say more about this towards the end of this book.

Today, more than seventy years after Bogolyubov's paper was published, the study of modulation instability is more vibrant than ever, and involves the efforts of research groups worldwide in the most diverse areas of physics; from optical fibres to fluid dynamics, from plasmas to quantum gases. Indeed, modulation instability was only experimentally observed in a Bose–Einstein condensate, the system studied in the paper published in 1947 by the Russian physicist, at the dawn of the third millennium.

The scientific successes achieved by Bogolyubov were definitely noteworthy; he received great acknowledgments both at home and at international level too. He was most likely one of the greatest theoretical physicists of the twentieth century. If he had given more weight to that minus sign, the pioneering investigation of modulation instability would most likely bear his name today. This episode shows how, frequently, the most important scientific discoveries are interwoven with details and coincidences. Even the greatest minds can neglect details that, in hindsight, contain the seed of decisive discoveries capable of influencing the scientific community's research and potential technological developments for decades to come.

NOTES

1. In the particular context of Bose–Einstein condensates this equation is called the Gross–Pitaevskii equation.
2. Note that, to be precise, technically it is a wave train – a series of waves propagating in the same direction with similar velocity and whose consecutive crests are equally spaced – that is destabilised by modulation instability in the fluid dynamics context.

CHAPTER 7

An Act of Creation

Nothing new under the Sun

ECCLESIASTES

The dinner was almost over, the tables crowded by fifty or so scientists already sated and clamouring, the musicians had stopped playing and gone outside to smoke, or maybe not, I do not remember very well, for the wine accompanying the various courses had been abundant and we were now drinking vodka. What I do remember, though, is a small man, not very tall and quite elderly, suddenly standing up, and that, seeing this, the room fell into silence. He raised his glass and called out, if my memory serves me well, something like this: "The universe is classic, but this place has been an act of creation!"

The man was Vladimir Zakharov and we were at Akademgorodok just outside the city of Novosibirsk, Siberia, Russia. Akademgorodok is an academic citadel, an ensemble of about forty research centres scattered among boulevards that twist and turn through pines and birches, boiling in summer, frozen and snow-covered in winter. Built in 1957, thanks to the initiative of the mathematician Mikhail Lavrentyev and of the President of the Council of Ministry Nikita Chruščëv, it was the

DOI: 10.1201/9781003462040-7

showpiece of Soviet Union scientific research and an oasis of intellectual and cultural freedom during the years of that regime. It was here that Vladimir Zakharov had worked since the 1970s, developing pioneering studies of modulation instability and nonlinear systems.

A few years earlier, as a PhD student, I had attended a physics summer school taking place at Villa del Grumello, on Lake Como. Zakharov was among the professors who were going to give lectures there. One of the main reasons that I attended was to listen to him speaking and to see him in person. The morning that his lecture was scheduled, the organisers announced that unfortunately Zakharov had not shown up, he had not even arrived at the hotel that was booked for him. I was hugely disappointed.

At the workshop in Novosibirsk some years later, Zakharov was not expected as either a speaker or an attendee. However, one morning, unannounced, he peeped out and sat on a chair in the front row of the room where the event was taking place. The speaker was talking about the theory of light turbulence. Suddenly Zakharov stood up, interrupted the speaker, and *motu proprio* went to the blackboard, wrote some equations, and started to explain how some results obtained some years ago by himself and by others could be relevant to the topic. He then had an exchange of views with the speaker, and finally sat down. Such improvised excursions were typical of him, and the scientific community had been accustomed to them for several years. Highly unconventional behaviour to say the least, and in a different context or from an ordinary scientist, it would surely be stigmatised.

For me, and I believe for anyone who researches waves developing in nonlinear systems, Zakharov is a keystone. He has made decisive contributions to plasma physics, to fibre optics physics, to turbulence theory, to meteorology, to fluid dynamics, and to mathematics. Along with various other acknowledgments, he has even had an asteroid named after him. He was also an established poet, and in a world where knowledge and expertise have become more and more fragmented and specialised, the fact that someone can still shine and significantly contribute to a wide range of disciplines gives me a great deal of hope. Zakharov has passed away only very recently aged eighty-four, and till the end of his life he has been an active researcher, continuously working on the study of modulation instability, and of rogue wave formation too. With his toast of "The universe is classic, but this place has been an act of creation!" Zakharov possibly wanted to say that the universe is ruled by inflexible laws like those of classical mechanics, which determine the future once and for all for the given initial conditions, but that sometimes we encounter completely unpredictable, astonishing, and extraordinary exceptions which we cannot easily explain using known schemes of thought. For him, maybe the birth of a space for thought and research like Akademgorodok constituted one of these exceptions. I don't know if this interpretation of mine corresponds with what Zakharov meant to say in that particular moment or to what he generally thought, but I like to imagine that the act of creation is like a flash of light which suddenly illuminates an anodyne moment of daily life; surprising, unpredictable and rare, but impossible to ignore. Perhaps it is made of the same elusive substance that a rogue wave is.

Rogue waves, also sometimes called freak waves, are extreme waves of huge dimensions, where by huge we mean statistically much taller than the average. They manifest themselves very rarely, "appearing from nothing and disappearing without leaving a trace" according to one of the verbal definitions most used by the physicists who study them. They are in the stories of sailors who, from boats at the mercy of the sea's fury, saw in front of them towering walls of water, sometimes over twenty metres high, which could destroy or swallow their ships. Rogue waves are rare events, meaning that the probability of one appearing at a certain time in a certain place on the sea's surface is extremely low compared to the probability that a wave with a small amplitude will form. Although its origins are in the oceans, the concept of rogue waves can be seen in other contexts too. For instance, it is customary to talk about optical rogue waves, namely particularly intense light waves which are produced when a light flow propagates along an optical fibre or while it bounces back and forth, imprisoned by of the mirrors of a laser cavity. In these latter cases, optical rogue waves are particularly intense flashes of light. These events are extremely luminous, rare, and apparently unpredictable.

Vladimir Zakharov was one of the advocates of the thesis according to which modulation instability, which we saw in the previous chapter, may play an important role in rogue wave formation. The theory that sees the origin of rogue waves in modulation instability says that a fluid wave oscillating at a certain frequency initially breaks up due to other wavelets oscillating at a different frequency, and whose amplitude is exponentially amplified. In what is called the linear phase of the instability, the initial wave becomes modulated by this new regular oscillation. However, when the oscillation reaches an amplitude comparable with that of the initial wave, a turbulent regime develops in the fluid. The turbulent regime is called the nonlinear phase of the instability. In it, the initial wave is replaced by a myriad of other waves called solitary waves, or *solitons*, which move in different directions and collide with each other.

Solitons are special waves that behave somewhat like particles. They are localised and not extended; in other words they occupy a very precise portion of space. They possess pronounced peaks, travel at a characteristic velocity, and their shape does not change upon propagation. After a collision with another wave a *soliton* will continue its journey unchanged, like a billiard ball which retains its characteristic shape after a collision with another ball. Solitary waves exist in various natural systems including the sea and in manmade canals, where they were first observed by the British engineer John Russell Scott. In his writings he gave the following description:

I was observing the motion of a boat which was rapidly drawn along a narrow channel by a pair of horses, when the boat suddenly stopped – not so the mass of water in the channel which it had put in motion; it accumulated round the prow of the vessel in a state of violent agitation, then suddenly leaving it behind, rolled forward with great velocity, assuming the form of a large solitary elevation, a rounded, smooth and well-defined heap of water, which continued its course along the channel apparently without change of form or diminution of speed. I followed it on horseback, and overtook it still rolling on at a rate of some eight or nine miles an hour, preserving its original figure some thirty feet long and a foot to a foot and a half in height. Its height gradually diminished, and after a chase of one or two miles I lost it in the windings of the channel. Such, in the month of August 1834, was my first chance interview with that singular and beautiful phenomenon which I have called the Wave of Translation...

Solitary waves can also be found in a wide range of contexts such as optical systems, where we find solitary waves of light which propagate without changing their shape for hundreds of kilometres along an optical fibre. In this particular scenario, solitary waves have been used, and to the best of my knowledge in some particular contexts still are, for transmitting information in commercial communication systems. Solitary waves can also be found in other physical systems including, but not limited to, plasmas, in optical resonators, in biological systems like proteins or DNA, in particular quantum gases at temperatures close to absolute zero, and in magnets.

But let us return to our ocean wave, the one that was just destroyed due to modulation instability, and has resulted in a myriad of solitary waves. The solitary waves that arose in the instability process propagate and collide in a chaotic and disordered fashion, and from these collisions, extreme waves can briefly arise. These have a much larger amplitude – perhaps ten times larger – than any of the individual solitary waves that form them. These extreme waves appear suddenly just after the collision process and then disappear without a trace. The *milieu* where *solitons* interact and collide constitutes the original matrix from which rogue waves originate.

This is one of the theories that aim at explaining the origins of rogue waves, and a vast amount of international research activity is dedicated to the comprehension of these probabilistically rare and frighteningly powerful phenomena. The great hope is that these efforts could be useful in the future to explain, and maybe also prevent, the formation of extreme ocean and atmospheric events. Once again, instability is the seed from which a dramatic qualitative change in a physical system grows, and it is instability that holds the power to generate new forms and phenomena like *solitons* and rogue waves.

CHAPTER 8

Bifurcations

Happy is the simplicity ignoring the bifurcations of doubt, but more wild and virile is the happiness which flourishes at the edges of the abysses.

ERNST JÜNGER

Imagine a wanderer walking across a flatland, who each day travels a certain distance. Although, step by step, small details around him change, on average the surrounding environment remains the same: some low vegetation, a few trees, the stones scattering the soil today similar to those of yesterday. One day, our wanderer comes to a point where the plain ends at the top of a cliff. Continuing forward the wanderer climbs down the cliff face and reaches a sandy beach. And look, all of sudden, everything around him is different: the sand under his feet, the rocky formations behind him, the surface of the sea in front of him. He finds himself in a new landscape, one of a different quality, and which is altered completely from where he was before. This transformation for our wanderer depends on the fact that he has moved, and that the region has a diverse landscape which at some points is completely different to others. The moment that

DOI: 10.1201/9781003462040-8

51

he crosses the cliff edge, a radical change manifests itself, and for as long as he stays below it the world is different for him.

The change that occurs when our wanderer transitions from the flatland to the cliff face constitutes a *bifurcation*. After he has crossed that border, everything is new to him: he must walk in a different way on the sand than he did on the hard soil; the reflection of the Sun on the water is not the same as it was on the dusty plain. The cliff edge is a threshold which discloses a new world. In the theory of dynamical systems, the landscape within which the wanderer has travelled in our picturesque image corresponds to what is called the *parameter space*. Every system is characterised by one or more parameters, each of which corresponds to an important feature. A parameter could be the temperature, or the energy injected into the system from the external world, the propensity of a system to lose or dissipate its energy, the interaction strength among its parts, and so on. The *parameter space* can in general be multidimensional; it can have as many dimensions as the number of parameters characterising the system. It defines the coordinates of the landscape describing all possible qualitative states of a system. It can be used as a map enabling us to orient ourselves and provide us guidance about the general behaviour of the system at issue: it tells us how to choose the parameters in order to obtain a certain behaviour or a different one. We can see the bifurcation as the junction point between two or more different regions of the landscape, like the change between different behaviours of the system as its parameters vary.

Instability processes are accompanied, as we have seen, by a qualitative change in the behaviour of the unstable system. The behaviour after the instability is very different from the one exhibited by the same system before it. So far, we have described how instabilities develop while assuming that the system at issue is already in a condition subject to instability. We have not asked ourselves *how* a system reaches a condition in which the system can go from being stable to unstable. Now we will explore exactly how a system *because of its history* can pass from a situation where it is stable to a situation where conditions are such that an instability is able develop.

Let us now look at some cases of bifurcation associated with instability processes. Consider for instance a laser. It consists of an optical cavity, typically composed of high reflectivity mirrors, which contains a material medium. The latter can be in a solid, liquid, or gaseous state, and is commonly called the active medium. A number of different substances can be used to constitute the active medium, for example ruby, erbium, helium and neon, carbon dioxide, as well as different semiconductors. Energy is pumped into the active medium from the outside, generally in the form of electricity. This makes it possible for some of the electrons that belong to the constituent atoms to make a quantum leap from a fundamental energy level to a level of higher energy called the excited state. At a certain point some of the excited atoms emit photons – light quanta – due to a phenomenon called spontaneous emission, and lose energy. As a result of this emission, the excited atoms return to the fundamental state. The emitted photons travel at the speed of light inside the cavity and bounce off the delimiting mirrors – unless they escape due to the non-perfect reflectivity of them – so that they pass through the active medium again. Their transit stimulates the emission of further photons by the atoms which are still excited: this process is called stimulated emission.[1] If the photon production occurs in a consistent fashion and overcomes the losses of light through the mirrors, this dynamic repeats until eventually a powerful light wave forms inside the optical cavity. The part of this wave which leaks out constitutes the laser light beam which we see manifesting in front of us. This, in essence, is the operational principle of a laser device.

Now, let us consider what happens when the laser is switched on. In this process there is a qualitative change in the system, a movement from a state where there are no photons in the cavity to a new state where the cavity hosts a great number of photons and the laser emits light. How are these two macroscopically and qualitatively different states connected to each other? Imagine now that the laser is off, and we start "pumping" some energy into the material medium. The laser remains off, and it does not produce any light. Small fluctuations of light intensity can arise like sparks inside the optical cavity perturbing the darkness, but they will rapidly decline; the "off state" is stable. If we gradually increase the flux of energy that we are injecting into the active medium, there will be a point at which we will see a sudden and radical change in the situation; now the laser emits a light beam with constant intensity. This qualitative change is called a *bifurcation* because, by gradually varying a system parameter – in this case the energy flux pumped into the optical cavity – the history of the system suddenly changes when the input reaches a critical value, as if the system were taking a different path. We call this the system's threshold value. In correspondence to the critical value, the "off state" of the laser becomes unstable. Small perturbations and fluctuations in the amplitude of the electric field, which are related to the number of photons circulating inside the cavity, are exponentially amplified and the light intensity transitions rapidly from a null to a large and constant value. The instability of the "off state" allows the *bifurcation* to occur, and a new state which consists of a light wave deriving its intensity from the energy stored in the excited atoms appears.

However, this is not the end of the story. Imagine that we continue to gradually increase the energy flux that we inject into the material medium of our laser. This small increase translates into a small increase of the light emitted by the laser. The laser remains in the same qualitative "on state", it is just that its light emission becomes more powerful. The "on state" is stable and robust; there are no qualitative changes. Small fluctuations of the light intensity, when they appear, are rapidly dampened. If we continue increasing the flux of injected energy, the intensity of the light emitted by the laser increases proportionally. But, luckily, nature is not so boring as to be satisfied with this. When the energy flux injected into the material medium reaches a second critical value, the kind of light emitted by the laser mutates radically. Small light fluctuations, small wavelets modulating this constant flow, are exponentially amplified: a new instability occurs. The stable "on state" becomes unstable and new scenarios are possible. The laser starts pulsating. The emitted light varies over time, reaching high intensity and then falling into the darkness, returning to high intensity and fading again, and so on and so forth. Depending on the particular kind of laser, these pulsations can be extremely regular, and when they are we can see them as flashes of light that all have the same intensity. These flashes repeat systematically every millionth of a second, or even faster. In other cases, the pulsations can be highly erratic and unpredictable; we call this regime optical chaos. The value of the control parameter which corresponds with this new change in the behaviour of our device, this new *bifurcation*, is called the laser's "second threshold".

Bifurcations and the qualitative changes associated with them manifest themselves in many different natural systems. As we have just seen in the example of the laser, in some cases the variation of the same control parameter can lead to multiple bifurcations, and an equal number of instabilities can be seen. A further example of bifurcation, which can be found in fluid dynamics, is the one giving rise to the convective phenomena associated with the Rayleigh–Bénard instability, which was discussed previously. Let us recall briefly what is it about. Consider a fluid located inside a container. Below the bottom of this container a heat source is present, and a lid covers the top of the container. The temperature of the container's bottom is higher than its lid. If the heat source has low temperature, the fluid remains in a steady rest state and its density and temperature remain roughly constant. But, if the temperature of the container bottom is raised above a critical value, a qualitative change occurs; the fluid is set in motion in a highly organised fashion with a kind of coordination and synchrony involving billions of molecules. Rotating currents are formed. A flux moves coherently from the container's bottom to its top, follows the underside of the container's lid, and then descends towards the bottom again, completing a cycle which then repeats. This flux constitutes a so-called Rayleigh–Bénard cell. Here a bifurcation also occurs when a control parameter, temperature, passes a certain critical threshold and the system's rest state become unstable. Instabilities are indeed the *trait-d'union* which in correspondence to the *bifurcation* carry the system exponentially from one state to a qualitatively different one. The instability of the old is what allows the birth of the new. In certain cases, in correspondence to the *bifurcation* point the system experiencing the change can acquire a possible state *A* or a possible state *B*. What determines its destiny at this turning point is chance, namely the particular form of the fluctuations. This kind of process can be seen in paradigmatic and simple form in the case of the Rayleigh–Bénard convection cells, in which path *A* consists of a cell rotating clockwise, and path *B* of a cell where the fluid rotates anti-clockwise.

Before leaving *bifurcations* behind, let us briefly return to our wanderer for a final look at the strange trails in the landscape of dynamical systems. First Scenario: standing on the beach, the wanderer turns his back on the sea and climbs back up the cliff. At the top, he finds himself back on the plain, and begins to walk towards back to his starting point, which he eventually reaches. This is what we would intuitively expect to happen in the world we encounter in our day-to-day experiences. If I move in one direction, I will reach a new place; if I turn around and retrace my steps, I will find myself back in the original place. Well, this seems to make sense. However, in many physical systems this is not what happens. Second Scenario: After the value of one of its parameters exceeds a given critical threshold, a system's behaviour suffers a *bifurcation*, transforms qualitatively, and acquires a form different from its original one. If the value of the control parameter is now gradually reduced, the behaviour the system exhibits on the way back is very different from its behaviour for the same identical value of the control parameter on the way there. This phenomenon is called hysteresis and teaches us that history and the past matter: there is memory in the most intimate physical structure of things. It is as if our wanderer, after retracing his steps, were to find himself on a mountain instead of the plain from which he set out. The system's behaviour does not only depend on the present value of its physical parameters, but on its history as well, on how it came to find itself exactly in that state. Reiterating the wanderer metaphor: it is as if we left home to buy bread, but though we took the road that led to the bakery, when we return home using that same road, we find ourselves in a completely different neighbourhood.

In the context of our daily experience, the idea of *bifurcation* refers intuitively to those events in life where a substantial and radical event manifests abruptly: an unexpected meeting that messes up our quite existence, a new path opening in front of us, circumstances inherited from the past crumbling before us and opening our eyes to new and unforeseen scenarios. In many cases we do not know what the key factor is, what the control parameter is, which, if properly modified, would allow us to realise or avoid big changes and transformations. Nor do we know if the big changes will be beneficial to us. Frequently, we suffer *bifurcations* without determining them. Maybe part of our existential effort towards what we love to call happiness or realisation consists of trying to understand what rules the *bifurcations* that rage through our biographies, in the hope of predicting and controlling them or, more modestly, of identifying them *a posteriori* to insert into the past a coherent and more reassuring narrative. Mapping *bifurcations* and understanding their rules would give us more freedom within our conditioned existence. This understanding might open the doors that get us ever closer to being the masters of our own biographies, or simply to better understand them.

I would be delighted to talk about these *bifurcations* with Dr. Yu Tsun, or with his ancestor Ts'ui Pên, who built in the mind of Jorge Luis Borges "*The garden of forking paths*" – a time-labyrinth – narrative mirror of the many-worlds interpretation of quantum mechanics proposed by physicist Hugh Everett. Its *bifurcations* occur in time: at the very moment when we perform any choice we give origin to a *bifurcation*, we create a possible future. However, in parallel worlds all the other futures relative to the possibilities we did not choose come into existence, including respective copies of ourselves. And these futures continuously multiply themselves in infinite worlds following new choices. A vertiginous attempt to account for the eluding physical meaning of the wave function and of its collapse on one of the possible states during the observation and measurement process of atomic and subatomic particles in quantum theory.[2] The bifurcations we have talked about so far describe instead the junctions, the trails, the available passages linking the possible ways of being of a certain physical system in *this* world. The bifurcation diagram of a system constitutes a map, an atlas, which can be consulted before deciding. It tells us which are the possible worlds that will be created in the course of time due to new worlds proliferation process ... provided that Everett was right and that, even if he was, these temporal quantum bifurcations survive the transition from the reign of the infinitely small to the one of macroscopic objects.

NOTES

1. The acronym LASER stands indeed for *Light Amplification by Stimulated Emission of Radiation*.
2. In quantum mechanics a particular physical system is described by a wave function containing all the information available on the system and whose evolution is ruled by the Schrödinger equation. The wave function describes the probability that the system – for instance a particle – has a certain value of a physically observable quantity (e.g. position, velocity, energy, rotation, etc.) if that quantity is measured. The measurement process of an observable physical quantity causes the wave function collapse from the indeterminacy of multiple possible values on a very precise value of that observable.

CHAPTER 9

Morphogenesis

I believe that this danger of the mathematician making mistakes is an unavoidable corollary of his power of sometimes hitting upon an entirely new method. This seems to be confirmed by the well known fact that the most reliable people will not usually hit upon really new methods.

ALAN TURING

On August 14th 1952 an article with title "The Chemical Basis of Morphogenesis" was published in *The Philosophical Transactions of the Royal Society of London*. The author, who would die just two years later, most likely by suicide after eating an apple poisoned with cyanide, was an English mathematician, and one of brightest scientific minds the twentieth century had known.

Alan Turing is well known to the general public for his contribution to the development of modern computers, having conceived the machine that bears his name, and for having cracked the Enigma code which the Germans used to exchange secret messages, a feat which made a significant contribution to the Allied campaign during WWII, or perhaps for his troubled personal affairs and persecution because of his homosexuality. Fewer people are aware of his fundamental contribution to biology and

DOI: 10.1201/9781003462040-9

to the theory of spontaneous genesis of forms in nature. But, to which forms are we referring here? We observe that the skin of some animals exhibits regular patterns likes stripes or spots. Furthermore, the physical forms of plants and animals have non-trivial structures and symmetries which change during the process of their growth. Turing wanted to explain how these forms and regularities arose in the first place, and how they developed subsequently. In his 1952 paper he studied mathematically a set of chemical substances that are responsible, proportionally to their concentration and density, for the development and differentiation of a living being's biological tissue. These substances, which he called *morphogens* and that in certain contexts correspond to particular proteins, can react with each other chemically and spread freely in spaces in which they are located.

Let us look at the two ingredients of the problem Turing studied. Firstly, the reaction. The chemicals involved in the reactions will eventually reach a state of equilibrium, in which the quantity of each involved reactant remains constant. As far as diffusion is concerned, this second ingredient considered by Turing can be visualised using the following images: gas particles which, having been injected into a room or a closed receptacle, spread erratically with random motion until they uniformly permeate the whole environment; or of a drop of ink which, falling in a glass of water, spreads in every direction until the liquid inside the glass is a uniform colour. Both the reaction and diffusion processes reach a state of uniformity, of homogeneity. By writing some equations describing a system where the processes interact with each other, Turing showed how a system in a state where the chemical components are uniformly distributed can become unstable. In other words, small disordered fluctuations in the systems, small non-homogeneities in the concentration of the substances, can be amplified to create spatial regions where their concentration is not homogeneous. Surprisingly, this non-homogeneity is not at all chaotic or random; instead, it exhibits highly organised structures, with clear regularities and symmetries.

The instability that Turing had discovered could generate forms with alternating stripes or hexagonal geometry as if by magic.

More than a century after Turing's paper on the spontaneous genesis of forms, the mechanism of regular *pattern* formation he mathematically predicted was observed in some particular chemical reactions. As far as biology is concerned, the situation is more controversial, and we cannot say that Turing's model constitutes a universal theory for the description of morphogenesis in biological organisms. There are, however, particular cases where Turing's model can contribute to explaining the formation of regular stripes on the skin of some fish species, the shape of protozoan structures, and mice mouth ridge pattern and toes spacing too. In these cases, instability becomes the vehicle of a qualitative transformation. It destroys one uniform order to establish a new one, which is differentiated, structured, and endowed with orientation and complex non-trivial symmetries.

It is important to stress that the forms which are generated through Turing instability are not imposed from outside; there is no one designing them. They develop spontaneously, shaped by the internal dynamics of the system. In a certain sense it is as if an orchestra has, without a conductor or any previous agreement between the musicians, given an harmonious and coordinated performance in which all the musicians spontaneously collaborate in an unplanned improvisation with no external direction. The morphogenesis described by Turing's model is an example of a spontaneous order emerging thanks to an instability. The instability of a particular state of the system, unleashed by minuscule fluctuations or non-homogeneities within it, becomes a force of destruction and at the same time a force which creates a new order. Turing's equations, despite being highly simplified models if compared to the remarkable complexity of the actual phenomena that they attempt to describe, or to the extremely realistic mathematical models that were developed later, have the merit of stimulating the search for an explanation of the generation of forms and *patterns* by isolating salient mechanisms of biological structures' growth and differentiation processes in certain cases.

Turing's legacy also has deep epistemological implications for the role and meaning of mathematical models in empirical science. Sometimes a simpler model with few ingredients, but the right and essential ones, can provide a better understanding of the causes and of the deep dynamics which govern a certain phenomenon than an extremely detailed model which appears to consider all the possible variables. This point is of particular significance where important decisions have to be taken on the basis of scientific predictions based on mathematical models. An extremely complicated mathematical or numerical model could predict some natural phenomena with utmost precision but leave the dominant causes of these predictions hidden amongst the crowd of all the possible variables. This could lead to the situation where the outputs of these models are taken as gospel without the underlying causes of them being fully understood.

Turing's mechanism considering a combination of a diffusive force, which tends to distribute a certain substance, and of a reactive force which tends to increase the concentration of the same substance in a particular region of space, has cut across the borders that separate different scientific disciplines. It has also revealed itself as crucial for explaining phenomena in research domains which are far removed from biology and chemistry. One field where Turing instability plays a fascinating and important role is ecology, particularly in the study of the spontaneous formation of the vegetation *patterns* which self-organise in semi-arid regions of our planet. Indeed, it is possible to explain the distribution of vegetation aggregates constituted by trees and bushes in desert regions through an instability process in the spatial distribution of the phytomass – i.e. of the organic matter constituting the plants. Another example can be seen in so-called *fairy circles*, zones of arid soil with a circular shape and diameters of about ten metres surrounded by rings of grassy vegetation, which can be found at the borders of deserts in Namibia and Australia.

Fairy circles are surprisingly long-lived and can last for several tens of years. Frequently, they arrange themselves so that they cover the soil in a compact and ordered way, almost like circular tiles that have been deliberately placed in close formation. The features of this kind of *pattern* can be estimated quantitatively using mathematical models which describe a reaction–diffusion and which spontaneously generate regular forms via Turing's instability.

Turing's idea has also been adapted by analogy to other physical systems such as optical resonators, which are studied by a discipline called nonlinear optics. An optical resonator is actually a physical system in which a light beam circulates repeatedly, and for a given time, along a closed path – as if in a circuit – which allows the selective propagation of certain light frequencies. An optical resonator can have various geometrical shapes and be made of various materials with which light interacts during its propagation. Typically, an intense light wave with constant amplitude is injected into the resonator. When this wave interacts with the matter inside the resonator, it can develop an instability. Small wavelets oscillating with a different frequency are exponentially amplified over multiple circuits inside the resonator until they reach an amplitude comparable to the one of the wave that was originally injected. At the end of the 1980s, in a pioneering study by Luigi Lugiato and René Lefever, the formal analogy was established between this optical system and the essential ingredients of Turing instability in biological systems. In the optical case, the reactive principle is constituted of a particular property of the matter of which the resonator is made, called nonlinearity, in virtue of which light tends to focalise and to increase its intensity at a certain point. The diffusive principle is connected to the fact that waves of different frequency tend to have propagation speeds that differ, leading to a light flux consisting of different waves spreading and scattering in the space through which it propagates. The latter effect is related to the phenomena of diffraction or chromatic dispersion.

The interaction between these two antagonist forces leads to the development of the optical Turing instability. As in the version seen in biology, in the optical scenario instability is the mediator of the manifestation of a new order in the system, consisting of regular geometric forms, true luminous crystals devoid of material structure but completely stable and robust: spatial energy structures. The forms which can be generated are numerous, but some of the most typical characteristics are hexagonal structures which repeat like a honeycomb, other like regular stripes where bands of light alternate with bands of darkness.

The concept of collective self-organisation, where thousands of distinct entities behave in a synchronous and coordinated fashion without an external influence, was not immediately accepted by the scientific community, where, as in other domains of human existence, intellectual conservatism and biased schematisations of experience can leave people blind to new evidence. This is illustrated by the story of the Russian chemist Belousov. At the beginning of the 1950s, Belousov observed experimentally that in a particular chemical reaction, collective oscillations manifest themselves in the behaviour of the reacting molecules. Namely, the entire solution became yellow then, after a few minutes, became colourless, then turned back to yellow; it continued to swing from one to the other for some time, like a sort of chemical pendulum. Belousov tried to publish a scientific paper in which he described his surprising discovery. However, important scientific journals in his discipline categorically refused to accept his results following the *peer review* process – the revision process carried out by experts in the field prior to publication. According to the experts of that time it was inconceivable and impossible that a phenomenon like the one observed by Belousov could even exist; there must be something wrong with his experiment. In the end the work was published in 1959, in a short format, in an anonymous medical bulletin of the proceedings of the institute where Belousov worked. It was only later that the scientific community accepted Belousov's discovery as a valid one, thanks to contributions by another chemist called Zhabotinsky, who resumed his works. The Belousov–Zhabotinsky reaction, as it was later named, became decisive for nonlinear science of the twentieth century, and was foundational for the development of *dissipative structures* theory and *Synergetics*, of which we will talk in a moment. Today, it is considered to be a textbook example, but Belousov did not live to see this glory; he died in 1970. In 1980 he received the Lenin prize posthumously. It is one of the highest honours awarded in the USSR for contributions to sciences, literature, and arts.

Research into self-organising processes received much of its initial drive from two great schools of scientific thought which were developed in Europe by Ilya Prigogine and Hermann Haken in the 1950s and 1960s. Ilya Prigogine developed the concept of *dissipative structures*. These are self-organising entities that exist thanks to a constant exchange of energy and matter with the surrounding environment. This exchange allows the *dissipative structure* to keep a high level of internal order contrasting the deadly entropy growth through its immersion in an open system, a system subject to fluxes communicating with the external world. The Turing *patterns* which form in biological systems are examples of *dissipative structures*, as are the analogous light structures which arise in an optical resonator, but a vortex in a turbulent system, or a convective cell in a fluid, are *dissipative structures* too. In a broader sense, highly complex entities like living beings could be considered to be *dissipative structures*. *Synergetics* describes complex systems as those that are constituted of multiple components reciprocally interacting, and which are able to self-organise and maintain salient features and order through a sort of cooperation between their constituent parts.

These powerful conceptual architectures tried to embrace with an ideal impulse, and with a coherent mathematical formalism as well, different disciplines like physics, biology, and chemistry, but also economics and sociology. Haken and Prigogine's efforts tried to account for the spontaneous formation of order in nature. The nineteenth century was the century of scientists like Clausius and Boltzmann and of the processes of the transformation of energy, work, heat, and entropy during the combustions inside mechanical engines, as well as of the meticulous statistical description of how swarms composed of an immense number of minuscule molecules, which form fluids and gases, contribute, through their dynamics, to collective and macroscopic properties of the latter. One of the of the century's scientific legacies is the thermodynamics of isolated systems: systems in thermal equilibrium with their environment, that is to say not undergoing constant fluxes of matter and energy in input and output. In these, the irreversible growth of disorder cannot be avoided. This is the content of the second law of thermodynamics which predicts that cosmic entropy can never diminish, it can only grow. The Universe is damned to stagger helplessly towards absolute disorder. Entropy is a measure of the disorder in a complex system. It is grounded by a two-level description of that system: the macroscopic one, which accounts for how the system appears to us when described by variables such as temperature, volume, pressure, and density of matter, and the microscopic one, which explains how molecules or particles of the system are arranged to give origin to the corresponding macroscopic state. A macroscopic state which can be realised by multiple configurations of its own parts, namely by many possible microscopic states, has more entropy than a macroscopic state that can be realised by few configurations of its constituent parts. If, by exchanging the positions or the properties of different system components we do not obtain significant changes of its qualitative macroscopic state, our system's entropy is high and therefore the system is disordered. Let us imagine two fluids, one red and the other blue, which have become perfectly mixed inside a container due to diffusion. If we swap the positions of various particles of different colours, nothing will change in the general landscape that appears before our eyes; we will always see a fluid that is basically purple. This is a system with high entropy. A container where the two fluids of different colour are separated, for instance where one fills the right half and the other the left, will have lower entropy. Here, we will see a qualitative change if we start exchanging the various particles' positions. If we continue the mixing process we will lose track of the original state. Once the two fluids are mixed the probability that they will fully separate in a spontaneous fashion due to the random motion of their constituent particles basically vanishes; the isolated systems cannot spontaneously transition from disorder to order.

However, in nature we observe processes where organisms differentiate and organise themselves to pass from a non-differentiated state towards a more differentiated one. During their lives, living organisms like plants and animals remain immune to disorder; they appear to contradict the second law of thermodynamics. During their evolution they even seem to acquire more differentiated and more ordered forms, for instance during the dynamics of an organisms' development from embryo. This observation led Erwin Schrödinger, one of the founding fathers of quantum mechanics, in a renowned essay called "*What is life?*" to introduce the concept of negative entropy – *negentropy* – to describe the process through which complex and living systems fight the spontaneous production of disorder ruled by the second law of thermodynamics. To preserve their order these systems must produce the opposite of entropy, namely, order. But mustn't entropy only increase? This apparent contradiction with the second law of thermodynamics is resolved by the fact that the entropy decrease always happens locally. It only happens in systems which are parts of the universe; it never concerns the entire universe. The salient feature of self-organising systems where entropy can decrease consists of the fact that these systems are open. This means that they constantly exchange matter and energy with the surrounding environment, they undergo input fluxes and dissipation; they are out of equilibrium. This enables them to increase their internal order even if the total sum of their entropy with that of the rest of the universe keeps increasing. Travelling towards thermal death, rivulets of order endure, pockets of *negentropy* sources, which become thinner and more perfect in the overall undifferentiated becoming: this is the future. Self-organised order generates itself on a large scale in such a way that the system exhibits emergent properties, i.e. properties which are not possessed by its constituent parts individually. Relationality among parts is the source of such forms of macroscopic order. The powerful paradigm of spontaneous self-organisation has multiple further ramifications; one of these is the self-assembly process of material particles with dimension of less than one millionth of a metre interacting among each other. In these systems, macroscopic structures can form called *soft* in virtue of their malleability with respect to external stimuli like pressure, temperature, and deformations. Examples of these ordered and complex material forms include gels, colloids, and liquid crystals, and they are especially present in the constitutive matter of living beings.

In our brief excursion through the amazing world of spontaneous morphogenesis and of the fight against the inevitable entropy growth, not only have we seen some important examples of self-organisation of order, but we have also probed a sort of universality of behaviour which manifests itself across distinct and apparently disconnected disciplines. A universality, or proximity, to which we will return more in detail towards the end of this book. But, before doing that, let us immerse ourselves in the infinitely small, and try to probe the stability of the fundamental components of matter, of the atoms and elementary particles that make up our entire Universe.

CHAPTER 10

Instabilities at the Atomic Scale

All composite things are impermanent,
They belong in the realm of birth and death;
When birth and death is transcended,
Absolute tranquillity is realised and blessed are we.

FROM A MAHĀYĀNA BUDDHISM SŪTRA

So far, our discussion of stability and instability has focussed on giant objects or processes like the orbits of planets or ocean waves, or on things that function on a human scale, like steam engines. We have also explored tiny objects, far smaller than us, like the waves of light forming in a laser or *patterns* self-organising in a chemical reaction, which are constituted by innumerable elementary components; photons in the case of light, or molecules of a given chemical substance in biological systems. In the following, we will go a step further in our journey and descend towards the infinitesimally small. We are about to pay a visit the atomic world and to interrogate the stability of elementary particles themselves!

At the dawn of modern physics, the understanding of the motion of macroscopic objects like planets and their orbits, or of free falling or rolling objects on inclined planes – think of Galileo's experiments on the tower of Pisa – were the focal point of physicists' interests. However, as the years rolled by, immersion into the unknown ocean of the infinitely small suddenly became possible, and with this, the investigation of the most intimate properties of matter. Between the end of the nineteenth and beginning of the twentieth century it became feasible to study experimentally what had previously been a theoretical and philosophical hypothesis, namely the existence of atoms. ἄτομος (*atomos*) in Greek means indivisible, and physicists at that time theorised atoms as the ultimate elements of matter, indivisible and immutable, which in different combinations and aggregates give origin to the diversity of material objects that surround us. This was an idea that had its origins in the distant past, first advanced by the ancient Greece philosophical school led by Democritus and Leucippus and subsequently resumed by Lucretius in Rome. The bold theory aimed at explaining why the objects we observe around us have different shapes, colours, and textures, that is to say distinct qualities, in terms of their intimate composition. The solution of the atomistic school was to postulate the existence of atoms, a set of fundamental, granular, minuscule, and indivisible entities, whose aggregation in different quantities and in different modes would explain the observable differences between objects through their microscopic composition. In this manner the atomists tried to account for the qualitative differences that we observe in objects through a quantitative principle.

It is curious to note, *en passant*, that the *clinamen*, the process that was theorised by atomist philosophers to explain the aggregation of atoms to form complex bodies, assumes the necessity of an effect evocative of the modern theory of instability. In it, a small perturbation is amplified, qualitatively modifying an original state in which atoms are separated, leading to a qualitatively novel state where atoms are arranged to form the things of the world. Indeed, atomists maintained that atoms were originally moving along in straight lines in a vacuum, and that small random deviations in direction slightly modified their routes, causing a series of collisions among them, which would eventually result in the formation of macroscopic objects.

The atomistic philosophical theory was widely revaluated in the nineteenth century, but this time it was supported not only by logic and abstract reasoning but also by empirical evidence. For instance, one of the proofs supporting the existence of the discontinuous and granular nature of matter was the explanation of Brownian motion by Albert Einstein. Brownian motion is the random and erratic movement of particles – in the case of its first observation, of pollens grains – when immersed in liquids or gasses. The features of this motion can be explained only by assuming that these particles interact through collision with moving corpuscles constituting the fluid or the gas. In the case of Brownian motion these corpuscles are the aggregations of atoms that we call molecules. This discovery strengthened the conviction that matter is constituted of discrete particles and is not a continuous substance. Other experimental work, in which solid-state crystalline materials were illuminated by x-rays, allowed scientists to reconstruct the mass distribution inside crystals based on the direction in which these rays were travelling after the interaction with matter. It was observed that mass was distributed in very precise zones of the space occupied by the crystal, corroborating the hypothesis that matter is constituted by discrete components. In crystals, atoms are distributed in an ordered fashion, unlike in liquids and gases where constituent particles move around and are disordered. In more recent times it has been possible to visualise single atoms using special instruments like the "scanning tunnelling microscope". However, research revealed that atoms were not the indivisible objects that they had been assumed to be. Rutherford's experiments, carried out at the beginning of the twentieth century, showed that the atom was not a monolithic structure. It was constituted by a nucleus, which occupied a small part of the atom's extension but comprised almost all of its mass, and of other particles distributed around this. Later it was discovered that the particles constituting the nucleus are protons and neutrons, while the particles "orbiting" around the nucleus are electrons. This non-indivisibility showed that the search for the ultimate constituents of matter was not over, and the necessity of continuing the search for further particles which were truly indivisible. The ultimate foundations of matter, so recently discovered, already needed to be redefined.

Several types of atom exist, and they are categorised by the number of protons contained in their nuclei. They are ordered and classified through the so-called periodic table of elements developed by Mendeleev. Many elements in the periodic table are unstable. This means that after a certain amount of time, the nucleus of the atom will disintegrate and transform into a different chemical element through the emission of particles and energy. This process is called decay and it is the basis of radioactivity. Some celebrated unstable elements are uranium, plutonium, radon, and polonium. The principal decay mechanisms are beta decay, where a neutron transforms, creating a proton, an electron, and an electron-antineutrino;[1] alpha decay, a process in which the atomic nucleus emits a particle consisting of two neutrons and two protons and which is in reality a nucleus of helium; and eventually gamma decay, consisting of the emission of photons – light quanta – following alpha or beta decays. The spontaneous decay of unstable nuclei constitutes the basis for natural radioactivity. We can push ourselves even further and ask whether these elementary particles themselves are indivisible and stable. The formidable theoretical and experimental effort of particle physics which has followed from these pioneering discoveries through the course of the twentieth century and beyond has been the attempt to provide an answer to this question: which particles are fundamental and not further separable into constituent sub-particles, and which ones can be further divided? For instance, the proton and the neutron, both essential constituents of each atomic nucleus, are in reality composed of other particles, called quarks. Particles can transform into other particles through the process of collision by being accelerated using extremely high energies. This happens in experiments occurring in huge accelerators like the Large Hadron Collider at CERN in Geneva.

But particles can also be intrinsically unstable. This means that they can spontaneously decay, generating other particles and emitting energy in the form of light. Imagine that we have a particle and we prevent its interaction with other particles. After a certain time we can no longer find it. However, in its place we find different particles. If we repeat the same experiment numerous times, we will see that the particle decays after different intervals of time. From the time that has to pass before a particle disintegrates in the various experiments, we obtain the so-called mean lifetime of the unstable particle. In general, unstable particles have distinct mean lifetimes. Take the neutron as an example, one of the three kinds of particles which are fundamental constituents of the atoms, together with proton and electron. As long as the neutron remains confined inside the atomic nucleus it is stable, but when it is outside of the nucleus it becomes unstable and decays after about fifteen minutes on average, giving origin to a proton, an electron, and to a third particle called electron-antineutrino. This is the aforementioned beta decay. However, other unstable particles exist too. We will not make a list here, but one example is the celebrated Higgs boson, which has recently penetrated popular culture. This particle, connected to the process necessary to confer mass to some of the existing massive particles, has an extremely short lifetime, equal to about 10^{-22} seconds (i.e. one ten-thousandth of a billionth of a billionth of a second).

These fundamental instabilities distance us from a vision of the universe where bigger and more complex forms and objects are made by small, immutable, and indestructible bricks, which are simply combined with each other to form the phenomenal reality. The same fundamental small bricks are, in reality, becoming entities, subject – under certain conditions – to instability, decay, and transformation. However, these fundamental instabilities are not only destructive. The products of unstable atomic nuclei decay are other, more stable nuclei, and the process can be considered to be a movement towards a greater stability. The fundamental constituents of matter do not always provide an ultimate anchor by which our certainties can be held. That anchor is at risk of destabilisation, being dragged into the vortex of change. We stand on intrinsically fragile terrain, which, if obsessively thought about, could bring us to the precipice of existentialist vertigo, of the possible impossibility of every possibility. This time not regarding ourselves as human beings but the same intimate material structure of the world. An equally radical and abyssal perspective.

NOTE

1. What we have generically called the beta decay is more precisely defined as beta –. There is also a decay process called beta + where a proton decays, emitting a neutron, a positron (a sort of electron that has a positive electrical charge called the electron antiparticle), and another particle called electron neutrino.

CHAPTER 11

Quantum Vacuum and the Origin of Perturbations

When Bodhidharma saw the Emperor Wu of the Liang
dynasty, the Emperor asked,
"What is the holy ultimate truth?"
Dharma answered, "It is Emptiness itself and there is
nothing holy."
"Who then is the one who at present stands confronting me?"
"I do not know!"

HEKIGAN-SHŪ (THE BLUE CLIFF COLLECTION)

Perturbations are an essential ingredient for the creation of instability. Without them systems in a situation of unstable equilibrium would stay there forever, no matter how precarious that equilibrium was. It is relatively easy to imagine the causes of perturbations in macroscopic systems. For instance, take a solar system, a star with some planets orbiting around it. In this case,

DOI: 10.1201/9781003462040-11

the presence of the gravitational field of another celestial body might cause a slight deviation in the orbit of a planet, kick-starting an instability. We can see a similar effect in manmade constructions such as bridges. Here the sources of perturbation are innumerable: the vibrations caused by vehicles in transit over the bridge, or those caused by people walking across it, or they could be those induced by atmospheric agents such as strong winds or waves. In the case of sea phenomena, we can imagine how small wavelets or ripples might be caused by the wind or by irregularities of the seabed, and might interact to generate bigger waves, giving rise to an instability process. Chemical solutions, where different substances react with each other, often contain residua of different substances or impurities which can perturb the regular course of the reaction. In lasers, it is the imperfections of the material elements constituting the optical cavity, like the mirrors or the active medium itself, which cause perturbations to begin in the light flux. Furthermore, mechanical vibrations or even temperature variations in the environment where the laser operates can be the source of small perturbations for the light waves forming inside the laser itself. As far as optical fibres are concerned, imperfections in their structure arising during fabrication, or from the bending of the fibre itself, could constitute perturbations for the light waves propagating through them. As in the case of lasers, they are also vulnerable to sound vibrations and fluctuations in temperature. Additionally, it is almost always the case that the laser or optical amplifiers used to prepare the waves for launch inside the fibre cannot synthesise, in an absolutely pure form, the desired frequencies. They add many oscillations to them, which have tiny energy variations at distinct frequencies, and are hence the source of possible instabilities. The term we use to describe these small fluctuations and undesired erratic variations is noise, and it is everywhere! It is present in every natural or human made system to a varying degree. Although noise is not necessarily always a problem, it is the seed from which perturbations grow.

But what concerns atomic and sub-atomic physics, and that which falls under the domain of the laws enunciated by quantum mechanics, is much more difficult to visualise. Here we must deal with a peculiar and paradoxical source of perturbation: the vacuum. Let us consider as an example the instability of an atom in an excited state. But, let's not get ahead of ourselves. What does it mean to say that an atom is in an excited state? Quantum mechanics tells us that the energy of an atom is constituted by discrete, "quantised" levels, i.e. an atom cannot have arbitrary energy level, only certain precise values which are a function of integer numbers called *quantum numbers*. The atom's energy is determined by which energy levels are occupied by electrons "orbiting" around the nucleus. The electrons fill the energy levels starting from the lowest position and increasing. When all the positions at the lowest energy level are filled by electrons, leaving no position unoccupied, we say that the atom is in the fundamental energy state. To simplify things, we can focus on the atom of the simplest chemical element, hydrogen. Hydrogen has a single electron "orbiting" around a nucleus constituted by a proton. When the electron occupies the lowest energy level, we say that the atom is in the fundamental state. To get the atom to transit from the fundamental energy state to a state that has greater energy it is necessary to supply energy to the atom from outside. This is achieved by subjecting it to an electromagnetic field. In this case a photon – a light quantum – can be absorbed by the atom, allowing the electron to jump to a higher energy level. In order for this process to happen it is necessary that the photon's energy corresponds to the energy difference between the atomic levels. Once the photon absorption process has taken place, we say that the atom is in an excited state, a state possessing a greater level of energy than the fundamental state. But is this change irreversible or can the atom somehow come back to the minimum energy state? The answer is that the process is reversible and that there are, in general, two ways for this to happen. If a new photon reaches the atom, and if its energy corresponds to the difference between the excited and the fundamental energy level, then it is likely that the atom will return to the fundamental state. In this process energy is emitted in form of light, and a second photon which has energy equal to the difference between the excited energy state and the fundamental one is produced. This process – in which a photon stimulates the emission of another identical photon, causing the atom de-excitation – is called "stimulated emission". It was first studied by Albert Einstein in a celebrated paper called "Zur Quantentheorie der Strahlung" ("On the Quantum Theory of Radiation"). Stimulated emission is the fundamental principle that underpins the operation of laser devices.

Incidentally, it is important to highlight that both the photon absorption process, with consequent atom excitation, and the stimulated emission process, with consequent emission of a photon (caused by the interaction of the excited atom with another photon), are of probabilistic nature. That is, if a photon possessing the right energy level to cause the excitation reaches the atom, there is a certain probability that excitation will take place, but excitation is not a certainty. Quantum mechanics explains how to calculate this probability. The knowledge of the probability with which the process occurs is the maximal knowledge we are allowed to possess in this context. The same holds for the stimulated emission process. We meet here one of the defining traits of quantum mechanics; the laws of quantum mechanics provide us with general information about the *probability* of a certain event occurring.

Forgive me for this digression; let us come back to our atom in the excited state. It would be natural to think that if we left the atom undisturbed in a space empty of all particles it would remain unchanged. We can imagine a perfect void for our solitary excited atom, empty of all matter and energy, and isolated from the external world: a hermitage in which it remains, preserving its energy. However, this isolation in an absolute vacuum does not produce the expected effect. After a certain characteristic time period (called the average lifetime of the excited state), the atom decays from the excited state to the fundamental one by emitting a photon that has an energy equal to the difference between the two levels. This process, which occurs without the stimulus of external forces or of other particles, is called "spontaneous emission".

Now, this process seems to contradict the paradigm according to which, in order to destabilise a state of equilibrium and trigger instability, the presence of a small perturbation is necessary. If the atom finds itself in a vacuum, there is nothing except the vacuum itself that could perturb it, but as a vacuum is a void, by definition it does not contain anything. We are faced with a sort of paradox. How can the vacuum generate a perturbation which triggers the instability of the excited atom?

The solution to this lies in an intimate understanding of the nature of vacuum revealed by quantum field theory. This theory was developed by combining quantum mechanics with special relativity theory. In it, fields, which are entities extended and permeating space, rather than particles, are the protagonists. Every field possess a kind of characteristic particle which describes its degree of excitation at a given point in space. In this sense particles are not fundamental quantities, as they describe the fields' degree of excitation or energy. At every point in space a field can exhibit different degrees of excitation, depending on the number of particles that are present at that point. If we consider the electromagnetic field as an example, its excitations, i.e. its particles, are the photons: the light quanta. Photons can be created or destroyed in interaction processes of the electromagnetic field with matter, like those we have seen in the processes of absorption and stimulated emission. Now, the vacuum state of the electromagnetic field corresponds to the situation where the field is not excited, i.e. where no particles are present. We would expect that, when all the field's excitations, namely the corresponding particles, are removed, the energy of this vacuum state would be equal to zero. But if we calculate the vacuum state's energy according to quantum theory, surprisingly we find a value which is greater than zero. This is vacuum energy, also called zero-point energy. The vacuum, according to quantum mechanics, is not a vacuum in the classical sense of the term. On the contrary, it possesses energy. This energy consists of fluctuations of the oscillating electromagnetic field. The minimum energy state, and we could say of maximal quiet, it is not at all quiet, but rather consists, if we want to provide a pictorial representation, of a sort of boiling and turbulent sea which constantly changes. Quantum vacuum fluctuations constitute the perturbations pushing the excited atoms to decay towards the fundamental state and emit light quanta: the vacuum's boiling is at the origin of the instability process of the excited atoms and hence of their spontaneous emission of photons.[1]

The vision of vacuum as a state in which energy is present and causes tangible effects requires a paradigm change with respect to the concept of vacuum as nothingness, as described by classical physics and one which, it seems, is much more rooted into common sense. What it is important to underline for the purpose of this book is the role of the quantum vacuum as the ultimate source of perturbation, and that it is its energy, with its relentless motion and the accompanying spontaneous decay processes at the atomic level, that underpins the most fundamental instabilities of nature.

Western civilisation traditionally abhors vacuum: *horror vacui* is, to a certain extent, one of its signatures. At the root of European civilisation, on sunny Greek islands, the search for the Principle began. It was Parmenides who made explicit the dominance of Being over Non-Being. Plato then perfected this afflatus by differentiating Being into these multiple essences which are the Ideas. From then onwards there has been a battle about how to interpret this Being and how it enters into relations with matter and becoming: of this consist the side notes to Plato about which Whitehead talks when referring to the entire *corpus* of Western philosophy. The absence of a foundation for Being has been stigmatised as conceptually fallacious and as the bearer of moral drift since the Sophists' time. Few, before the postmodern *Weak Thought* have ennobled the absence of a stable foundation of Truth, Subject, and History.

But there are exceptions. If postmodernism seems to define a *weak Being*, incapable of ontologically founding Truth, then Meister Eckhart's philosophical-mystical thought, through its identification of the bottom of the soul meant as Nothingness and God itself devoid of any positively definable attribute – in the moment in which the subject through detachment recognises himself as devoid of substantial and autonomous reality – is maybe one of the rare celebrations of the vacuum as the cornerstone of metaphysics.[2]

However, *Horror vacui* does not remain confined to the minds of metaphysicists. It becomes flesh in the everyday lives of those who are unaware of sophisticated intellectual debates. The waves of psychic apprehension that travel like shivers up the backs of the Western masses require background noise, the bustle of trivial and quick entertainments, to drive out the fear of silence. Vacuums and silence are spurned and often filled with whatever is within arm's reach. In the ancient East, the Vacuum had the dignity of foundation. Emptiness is the fundamental essence of reality and the ultimate spiritual state which Buddhist meditative practice cultivates. Regardless of whether its direct experience is achievable through contemplation, emptiness in Buddhism can be superficially philosophically summarised in the concept of the co-dependent generation of things, according to which each entity does not exist independently and in isolation, but only in virtue and as a function of its interactions and relations with all other entities – things are devoid of independent self-sufficient existence. This existence purely founded on the relations between objects and even elementary particles also emerges from the so-called *relational interpretation* of quantum mechanics (although in a different context to our discourse on vacuum and its fluctuations). According to this interpretation, as Rovelli puts it, "the properties of an object *are* the way in which it acts upon other objects; reality is this web of interactions."

From the end of the twentieth century, one of the most sophisticated formulations of modern science, quantum field theory, has revealed that the vacuum is a positive protagonist, an active and rationally accessible one. Quantum fields distributed in space-time are the fundamental entities, and their vacuum state is the matrix from which particles are born and where they die. The vacuum state is active and endowed with positive and ontologically definable attributes. A place, both of Being and of its deprivation. Who knows, maybe along this speculative path, a reconciliation could be sought, which would overcome the antithesis between East and West, between silence and noise, between presence and absence.

NOTES

1. The probability per unit time that an excited atom spontaneously decays emitting a photon is described by a term called "A Einstein coefficient". To be precise, vacuum fluctuations give a contribution to this probability equal to half of A; the remaining part derives from a different effect called "radiation reaction" (for more details see for example the book by P. W. Milonni cited in the bibliography).

2. The one by Eckhart is an example of negative theology, namely of that way which aims at defining the Divine through negation only, and not by affirming its qualities, i.e. it says only what God is not.

CHAPTER 12

Instabilities in the Head

Each one's destiny: instability

ATTRIBUTED TO PHILIP ROTH

We have travelled along a path which has brought us from planetary phenomena to the most intimate atomic structure of the cosmos with instability as our guiding star. The quantum processes we saw in the previous chapter seem like impalpable abstractions of thought, their content far removed from our everyday thought processes. But now it is time for us to think about thought itself.

Thought processes are supported by the physical correlate of brain activity. The latter is based upon thousands and thousands of electric discharges every second which propagate in our brain via oblong cells called neurons. The electrical signals which cross the neurons are transmitted to the muscles to control motor stimuli. They also travel to the brain from receptors located on the skin to convey information connected to perception. The single neuron is the fundamental biological unit of the nervous system and is constituted of a body called the soma, possessing a prolongation called the axon, which is used to send electrical

DOI: 10.1201/9781003462040-12

signals to other cells. Around the soma, filaments and ramifications called dendrites receive electrical signals from surrounding cells. The signals are transmitted from one neuron to the next in the form of electrical impulses called action potentials, which can travel at speeds of up to 400 kilometres per hour. In a rest situation the difference in electric potential between the interior and the exterior of the neuron's body is constant, and it is called the equilibrium potential. This electrical potential difference is determined by the fact that electrically negatively charged ions can easily penetrate the cell membrane, while positively charged ones remain outside. The electrical potential that is formed in this way is negative, i.e. it is lower inside the cell. Our neuron's dendrites receive information from other neurons or surrounding cells in the form of chemical substances secreted by the latter and called neurotransmitters. The messages that the neuron receives are integrated and combined in the dendrites, leading to a modification of the electrical potential between the interior and the exterior of the neuron itself. This can give an inhibitory contribution, that strengthens the rest state of the neuron, or an excitatory one that stimulates its activity. If the signal is intense enough to cause the electrical potential difference between interior and exterior of the neuron to vary substantially, in excess of a particular critical value, a dramatic change in the behaviour of the neuron takes place. The rest state becomes unstable, small perturbations are exponentially amplified, and an electrical impulse – called *spike* – is generated, which propagates from the membrane and along the whole axon, where, at its extremity, it will give origin to the emission of new neurotransmitters, continuing the chain of communication with other surrounding cells.

So, in summary, while it may appear unusual and counterintuitive, we even find an instability underpinning the operation of the brain and of the nervous system. In this context, the instability control parameter is the electrical potential difference between the interior and the exterior of the membrane, which has to be above a threshold value in order to unleash the neuron response. Above this threshold a bifurcation occurs; the equilibrium state in which the potential across the membrane that was constant becomes unstable and the neuron's behaviour changes qualitatively as the latter starts to emit electrical impulses. After generating the signal, the neuron needs a certain time to reconstitute the equilibrium electrical potential; this is called the *refractory period*. Once equilibrium is reached again, the arrival of a further signal from a neighbouring neuron can stimulate the repetition of the entire process. Generally, the impulses emitted by the neuron all have the same intensity regardless of the magnitude of the stimulus, but the latter has to be larger than a certain critical value for the entire process to take place. Indeed, neurons respond, or "fire" to use the jargon, at their maximum capacity or they do not "fire" at all; they are all or nothing. The instability of a neuron's rest state can result in various bifurcations which give rise to different types of electric impulse. In some cases a single impulse is formed, in others a regular train of impulses, and in others still the electrical impulses are generated in groups (the latter phenomenon is called *bursting*).

Incidentally, it is interesting to note that the excitability phenomenon described in this chapter is not a behaviour exclusive of neurons. It can be found in other contexts too, for instance in a certain kind of laser exhibiting the emission of giant light pulses in correspondence to particular perturbations and external stimuli. Mathematically this dynamic is analogous to that of biological neurons, making these lasers a sort of optical neurons. The imitation of the computation and information manipulation processes seen in the biological neurons of the nervous system in optical devices has given origin to a new discipline called *neuromorphic photonics*. This discipline involves a great number of optical devices including excitable lasers, and aims at developing a new generation of computers which employ functional architectures which differ substantially from those on which the computers we know are based by using light instead of electricity to perform calculations. With the ambition of overcoming the Von-Neumann computational paradigm (according to which memory and computation unit – or processor – are separated entities) on which existing computers are built, huge efforts from both academia and industry currently aim to revolutionise the *status quo* and develop new photonic machines where memory and computation are not separate processes at the *hardware* level. This should increase computational power and at the same time reduce energy consumption through the integration of optical processes, in particular miniaturised photonic chips.

We can summarise as follows: The generation of the action potential in a neuron and its transmission through the latter determine the atomic event constituting the indivisible and fundamental brick, the *quantum* – if we want to borrow from physics a term to use in our language – of the entire nervous system and of all brain processes activity, hence ultimately of human life itself. This *quantum* is grounded on an instability. It is this instability which enables the development of complex and fundamental tasks connected to perception, movement, and survival of the human being, which enables the sophisticated activities of philosophical and scientific thought, or of artistic creation.

CHAPTER **13**

Pillars of Hercules

One shore and the other I saw as far as Spain,
Morocco, the island of Sardegna,
and other islands set into that sea.
I and my shipmates had grown old and slow
before we reached the narrow strait
where Hercules marked off the limits,
warning all men to go no farther...

DANTE ALIGHIERI, INFERNO *(CANTO XXVI)*

As we have reached the description of the instability associated with our thought processes anyway, let us remain on the theme of abstraction, and let me indulge myself a little by discussing some general considerations that concern all of the instability phenomena that we have seen so far. I would like to talk about the limitations of the mathematical techniques that are used to analyse stability, and about how these limitations affect our capability of knowing nature. In order to fully understand these limitations, we must say a few words about the techniques that allow us to study the stability of a system.

The mathematical technique of linear stability analysis first used by Maxwell in the study of Saturn's rings and of governors has subsequently been extended to many other systems. It has been modified, where needed, in order to account for a range of phenomena. It is a formidable tool, allowing us to calculate whether the perturbations which disturb a given system will be dampened, will remain steady, or will grow unbounded. If they will grow, stability analysis tells us how fast and in which way they will do it, for instance by oscillating or in a continuous and uniform fashion. However, this mathematical precision comes at a price. Linear stability analysis uses substantial simplifications in the equations that rule the dynamical evolution of perturbations. In particular, in obtaining the equations describing the perturbations, some terms are neglected which are unnecessary when perturbations are very small. These neglected terms, called nonlinear terms, contain the necessary information to describe the perturbations when the latter's magnitude has become comparable with that of the state of the system whose stability one wants to study. But, in their presence the equations cannot be exactly solved in general. We are, then, facing the Pillars of Hercules of stability analysis. Either we neglect some terms in the equations and obtain a solution valid only while the perturbations remain small, or we keep them, making the equations valid when the perturbations have been abundantly amplified, but in general unsolvable. We can only know, in a mathematically exact way, the instability dynamics in its initial phase. However, there is a way to quench our thirst for knowledge and to cross, with a certain justified *hubris*, the limits of our analytical abilities. Let me demonstrate this point through an historical anecdote.

In the early 1950s, Enrico Fermi, John Pasta, Stanislaw Ulam, and Mary Tsingou were studying a physics problem that they could not solve it mathematically in an exact way. They decided to entrust the solution to a computer; it was the first numerical experiment in history. The computer was at Los Alamos in the United States and was called MANIAC I. It weighted half a ton and had been used to execute calculations which led to the construction of the hydrogen bomb. The problem that our scientists studied with the aid of this electronic computer was not apparently directly connected to the study of instabilities. It had to do with the investigation of whether, in a system possessing different modes of oscillation,[1] the energy initially concentrated in a particular mode would distribute equally to all the remaining ones, or whether it would remain confined to one or a few particular modes. Their intuition was that the energy would distribute and that the system, in the long term, would exhibit oscillations with various frequencies. However, this numerical experiment showed that, although the energy was initially transmitted to other modes of oscillation, after a while it came back to the initial one. This returning was baptized *recurrence*. Indeed, this term was transferred to the domain of instabilities to signify the returning of energy to the primary wave after it has been ceded to the perturbation waves, as we have hinted before. Important common traits between the phenomenon studied mathematically by Fermi, Pasta, Ulam, and Tsingou and modulation instability are the fact that they both occur in nonlinear systems and that they are associated with the dynamics of solitary waves. The results on the *recurrence* phenomenon, described on an internal laboratory report and not published in any scientific journal, have constituted the incipit for the birth of nonlinear systems science. The name of Mary Tsingou did not appear among the authors though: "Report written by Fermi, Pasta, and Ulam. Work done by Fermi, Pasta, Ulam, and Tsingou" one can read at the beginning of the document. This ensured that when people referred to those results in the specialist literature, they never made any reference to Mary Tsingou, neglecting a notable scientific merit. One has to consider that at that time writing a computer program was not the relatively accessible task it is today. It was extremely arduous and required formidable ingenuity and creativity. It is only very recently that the contribution of the woman who actually wrote the code necessary to carry out the first numerical experiment in history has been officially recognised by the scientific community.

Since that pioneering attempt, the use of computers in the solution of mathematical equations has become an essential part of scientists' research armoury. With the computer it is possible to perform true numerical experiments which reveal the behaviour of complex systems even when it proves impossible to deal with the problems through mathematical analysis. Nowadays, the use of computer simulations is a usual and established practice in almost all scientific disciplines. These simulations are used to predict the behaviour of complex systems and to prepare experiments in order to maximize success. Alternatively, they are employed to verify the plausible validity of mathematical models and theories before performing a real experiment.

Let us get back to the study of instability dynamics, which is no exception and today takes place through the use of numerical computer simulations which unveil the whole dynamics by algorithmically solving the equations. A huge part of scientific literature is devoted to developing algorithms, allowing the solution of the most diverse mathematical equations, in the most precise and fastest possible way (despite this, in certain particular problems, computer simulations can take several days or even weeks). Strictly speaking, solutions obtained using the computer are approximated too, and their validity is within a certain tolerance margin which depends on the algorithms employed. However, this is very often sufficient to provide us with all the information we need to know the advanced stages of instability processes.

The formation of self-organising processes and *patterns* or *dissipative structures* are knowable theoretically only thanks to calculations performed by computers. For instance, the dynamics involved in the generation of turbulence and of its whirling coherent structures can be described mathematically only in statistical terms, and for an effective visualisation of the phenomenon one has to resort to computer simulations. The same can be said of a large part of hydrodynamics' instabilities and of the light crystals' formation in lasers. Informatics comes providentially to help us, and to illuminates the path in the decisive moment when instabilities open wide the doors that can lead to systems' transition to a new order.

However, paradoxically, with a last-ditch effort instability can influence the informatics algorithms that we use to investigate the instability of natural systems. Indeed, these algorithms have to be designed and implemented in such a way that the numerical approximations and the computational errors unavoidably associated with their operation stay small and do not grow exponentially, especially since in this last case they would compromise the convergency towards the correct result of the informatics calculation. When numerical computation errors are exponentially amplified an algorithmic instability takes place, variables can assume huge values or ones so small that they crash into the precision limit of electronic computers; the computational process explodes and stops. This circularity reminds us once again of the ubiquitous and transversal nature of instability processes; in certain limit cases instability may concern both the process under investigation and the tools used to study it.

Algorithmic instabilities, if considered as being limited to the example just mentioned, can appear as mere academic curiosities. However, the scenario dramatically changes if we bring to our attention the already vital role played by *software* and algorithms in almost all aspects of our daily lives. Algorithms are woven into commercial and financial operations, into the credit supply process, into dating applications, into the operation of social networks, into the control of production and supply chains, into the formation of political consensus, into the profiling and identification of citizens by the states. Their stability or instability becomes vital to keep in place or to destroy the processes which exist thanks to them: the existential salience of algorithmic instabilities is greater than we think, and it is destined only to grow.

NOTE

1. Each physical system possess', in general, different characteristic frequencies, also called *normal modes*, with which it can oscillate.

CHAPTER 14

Exponential Nature

Anyone who believes exponential growth can go on forever in a finite world is either a madman or an economist.

KENNETH EWART BOULDING (ECONOMIST)

Our intuition about the future is linear. But the reality of information technology is exponential, and that makes a profound difference. If I take 30 steps linearly, I get to 30. If I take 30 steps exponentially, I get to a billion.

RAY KURZWEIL (FUTUROLOGIST)

A further consideration I would like to present is connected to the way instabilities manifest themselves and in which perturbations are amplified. Instabilities are processes that occur on exponential temporal scales. Every instability process where tiny perturbations are amplified has a growth rate dependent on the particular parameters of the system in which it takes place. This growth rate can be mathematically calculated using the stability analysis. Let us be clear though; the characteristic time over which an instability develops can be extremely long in human terms, for example, consider the time frame of the instability of our solar system that we dealt with at the beginning of this book.

DOI: 10.1201/9781003462040-14

What matters, however, is the following qualitative picture: after the initial burst, in an unstable system, perturbations' growth does not continue for very long. In order to grow, perturbations require material and energetic resources from which to draw in their amplification process. These resources can already be present in the system or can be received from the external world, and absorbed in the initial phase or during the instability process. After this, growth becomes slower and a saturation process follows where the exponential dynamics ends. The system can transit towards a new static equilibrium, towards a regularly oscillating dynamics, or towards a chaotic and turbulent one. As we have seen, in certain cases an inverse process is also possible, where the perturbations' energy temporarily returns to the form it had before the instability. The fact is that finite resources constitute an important brake to powerful exponential growth in instability processes. Extreme rapidity pays the price of a limited duration, like the short life of radiant glory of Homer's Achilles.

So far, we have met a series of examples that show us how instabilities subtend to qualitative radical changes in the behaviour and form of the systems where they occur; remember the *pattern* formation on animals' skin, the generation of optical pulses in lasers, the appearance of vortexes in turbulent structures? As they are associated with instability processes, these changes occur with exponential timings once the particular system at issue finds itself in the conditions under which instability can happen. Instability initiated change is sudden and upsetting; however, the exponential process has a finite duration and eventually transforms into a new *status quo*. The transition cannot go on forever, and the system needs to readjust and "rest" at a certain point. The lesson that we learn from this is that the exponential processes we see in unstable dynamics are never a free lunch; they are sustained by existing resources which have to be provided constantly to feed them. Their sustainability and repeatability are intimately connected to the context in which they occur and to the availability of the necessary ingredients.

Exponential processes in an unstable system are, in reality, small parts, fragments of a much longer history. They are moments which are transformative, but limited in time and by boundary conditions.

However, a vertiginous landscape becomes open for contemplation if we broaden our horizons. We can do this by venturing into the rough terrain of speculation, and mentioning the position of the inventor, "futurologist", and engineer at Google, Ray Kurzweil. A few years ago, he came talk to us physicists and photonic engineers as an exceptional guest at an important conference on laser science and technology, which was taking place between spring and summer in San José in the Silicon Valley. Kurzweil maintains that an exponential growth could, in certain contexts, sustain itself and continue to accelerate. This would occur if the change processes that manifest in a system were to act in a certain way on themselves, modifying and strengthening themselves as a direct result of the outcomes of their own change processes. Kurzweil is proposing sort of a *feedback loop* which could lead to a continuous acceleration of change itself. The resources necessary for the next stage of the growth process would be extracted from progresses of the previous growth process. Ray Kurzweil claims, and seems to be able to prove, that this exponential dynamics has manifested itself in the technological innovations seen in recent centuries: the computational capacity of machines; telecommunications; miniaturisation; DNA sequencing, but also in other contexts, such as the biological evolution of life on planet Earth from its beginning until the present. Kurzweil argues that these processes are ruled by the so-called *Law of Accelerating Returns*, which in simplified terms we can represent by analogy as the growth dynamics of capital through compound interest in financial investments.

To see how this operates in our own societies, we can think of how the construction of machines and rudimentary technologies has shortened the time that has been needed for the development of more complex machines and technologies. The advent of more complex machines and technologies allowed the creation of computers, which in turn accelerated the creation of even faster and more effective machines and technologies, and of new high-performance computers with radically improved efficiency. This race, which also embraces other technologies like robotics and genetics, could reach the Singularity during twenty-first century; a technological change so radical as to determine a "rift" in the fabric of human history and where there could be, among other things, a fusion between human and computer created intelligence. A component of this exponential growth process is exemplified by the so-called empirical law by Gordon Moore, Intel co-founder, which describes the exponential increment of the number of transistors present in an integrated circuit, which was observed from the 1960s onwards: the number of transistors doubles every year. Although Kurzweil admits that a single exponential process can become exhausted due to finite resources or due to other limitations – as is the case of integrated circuit technology, where the transistor dimension approaches the atomic – the exhausted process would be replaced by an even more efficient one, which would continue the global trend initiated by its predecessor. This would be seen by the fact that computational capacity and the performance of computers grew exponentially during the twentieth century, but that this growth was supported by different processes or paradigms, each one of which has contributed to the exponential growth; thermionic valves were followed by individual transistors and then by integrated circuits. Each one of these has been, or will be, saturated eventually and been succeeded by a new paradigm.

I cannot say precisely whether this Promethean vision could somehow describe the continuous, unbounded growth of instability processes. In the case of single, simple systems, like the ones we have described in these pages, I would say no. It is *possible* that in a broader macroscopic system, where various unstable subsystems interact, diverse instability processes could concatenate like the stairs of an infinite staircase.

Staying in the terrain of speculation, I would say that I struggle to conceive of this type of exponential process as absolutely unstoppable. Sometimes I fantasise, tickled by the potential of infinite growth, but more frequently, I have to admit, I find myself more at ease with finitude, or as Alessandro Baricco puts in the mouth of Novecento, Danny Boodman T.D. Lemon:

> Now think: a piano. Keys begin. Keys end. You know that they are eighty-eight, no one can fool you about this. They are not infinite. You are infinite, and inside those keys, infinite is the music you can play. They are eighty-eight. You are infinite. This I like. This one can live. ... If that keyboard is infinite, then... On that keyboard there is no music that you can play. You are sitting on the wrong stool: that is the piano where God plays.

Universal Forms

If Reason and True Opinion are two distinct Kinds, most certainly these self-subsisting Forms do exist, imperceptible by our senses, and objects of Reason only; whereas if, as appears to some, True Opinion differs in naught from Reason, then, on the contrary, all the things which we perceive by our bodily senses must be judged to be most stable.

PLATO, TIMAEUS

We have briskly crossed the landscape of perils and of challenges to received thought which have characterised the articulation and development of the concept of instability in modern science, throwing our gaze here and there on some particular examples. Instabilities arise everywhere, like a hydra whose many heads peep out from the thicket of nature's entangled plot. In the detailed multiplicity of the different systems we have explored – quantum gases, lasers, optical fibres, planetary systems, chemical reactions, cerebral cells, arabesques, and the drawings on animals' backs – there is something that endures, a residuum, a red thread that somehow unifies the various phenomena. It identifies their similarity, a sort of solid skeleton distributed through this plethora of apparently disconnected situations and appearances.

DOI: 10.1201/9781003462040-15

If we examine this red thread more closely, we see that it is in turn constituted by widely differing components. Now, I would like to show a couple of them to you.

The first component that enables the identification of common traits is the mathematical language generally used in the description of natural phenomena. It pertains and adapts to the description of a variety of systems, independently of the particular material substrate. Salient features of mathematical language used in the different fields of knowledge are the evolution law of a physical, chemical, engineering, or biological system which can be written in the form of one or more differential equations. In many cases it is possible to find a solution to these equations which corresponds to one particular state of the system at issue. One can apply the stability analysis of the solutions found and characterise the evolution of the small perturbations which disturb such a state to understand how they behave; whether or not they grow, and if they do it how they do it. During our brief journey, this mathematical language and these universal techniques have been described in an approximate way, as far as the everyday and non-technical language allows it, making an abundant use of similes and metaphors, and identifying analogies among the behaviours of the various systems. The evolution of a process, the particular state of the system whose stability one wants to test, the bifurcation in the system's "history", the small perturbations which, from infinitesimal magnitude, grow exponentially.

This universality of language is a fundamental problem of the philosophy of science. How is it possible that mathematical language can describe natural phenomena, and in particular the ones we encounter in physics? This question has been open for debate since Galileo stated in *Il Saggiatore* that

> Philosophy is written in that great book which ever is before our eyes – I mean the universe – but we cannot understand it if we do not first learn the language and grasp the symbols in which it is written. The book is

written in mathematical language, and the symbols are triangles, circles and other geometrical figures, without whose help it is impossible to comprehend a single word of it; without which one wanders in vain through a dark labyrinth.

On this matter various thinkers, scientists, and philosophers of science have expressed their opinion. It does not seem to me that we have come to a convincing conclusion. It seems to me that a statement of physicist Eugene Wigner is extremely relevant here. He says that

the miracle of the appropriateness of the language of mathematics for the formulation of the laws of physics is a wonderful gift which we neither understand nor deserve" and that "the enormous usefulness of mathematics in the natural sciences is something bordering on the mysterious and that there is no rational explanation for it.

Another component of our red thread is the presence, among the equations of motion describing the evolution of different natural systems, of classes of equations which we can define as universal forms. Indeed, the evolution of systems that are completely different to each other can frequently be described by almost identical – or, at least, very similar – mathematical equations. For example, in the cases of the propagation of light waves in an optical fibre and of water waves developing in oceans where the depth is substantial, or of those manifesting in Bose–Einstein condensates. All these systems are described by a mathematical formula called the nonlinear Schrödinger equation. Elsewhere, certain types of lasers, fluids in convective motion, phenomena connected to superconductivity and superfluidity, the behaviour of liquid crystals in particular contexts, and also some processes of string theory in particle physics, are all described by a mathematical model called the Ginzburg–Landau equation. This ensures that when we formulate a prediction on a certain dynamical behaviour or phenomenon, for instance an instability or the existence of a wave having a particular shape, using a universal equation, we can expect that this phenomenon manifests itself in all the different systems described by the equation. These universal equations constitute a fundamental architrave of scientific thought, and the hidden soul of the phenomenal structure in a range of different fields that seem unconnected at first sight. Prigogine and Haken's attempts with their respective theories of *dissipative structures* and *Synergetics*, at which we have hinted, go somewhat in this direction. They try to identify the common structures and processes underpinning the instabilities which are connected to self-organisation processes.

If the Platonic view of the world introduced ideas with the purpose of accounting for what, endowed with superior ontological quality, remains eternal and stable through the turbulent flow of becoming, I like to think that these universal mathematical structures, these ubiquitous equations, are made of the same stuff somehow, of Platonic ideas. They are maybe ideas regulating becoming, structures which are more general with respect to single things ideas, forms ruling universal change processes which are materially embodied in instances and in very different and particular systems.

CHAPTER 16

Riding the Tiger

Necessity is an evil, but it is not necessary to live in necessity.

EPICURUS

We are approaching the conclusion of this book and after the flight of fancy we have took in the previous chapter, it is time to place our feet back firmly on the ground. Beyond the desire to expand human knowledge and its conceptual beauty, one may have good reasons to ask if the study of instabilities has any practical and tangible application. Maybe, the importance of the deep understanding of instabilities which enabled the development of reliable feedback mechanisms, as in the case of the governors developed during Industrial Revolution that we talked about at the beginning of this book, isn't quite so eye-catching today. In reality, feedback mechanisms are widely used in modern technologies, and in particular they are an essential component in numerous electronic circuits; furthermore, they play a fundamental role for instance in the operation of self-driving vehicles, as well as in robotics. From this point of view, certainly, the study of instabilities doesn't lack practical significance. However, there are several further contemporary cases where knowing about a system's instability features is extremely useful. In this chapter

DOI: 10.1201/9781003462040-16

I want to reflect on the usefulness of instability, but I also want to examine the trust we place in it. If we were told that a bridge was precarious and unstable, would we cross it in our car? I don't think we would. This is instability to be avoided. Are we willing to trust surgical instruments, clocks, or communication systems whose operational principles are based on instability?

Humankind has always feared the terrible rage of uncontrolled natural forces. Earthquakes, hurricanes, floods, storms, landslides, tsunamis, avalanches, volcanic eruptions; these are only some of the arrows that nature has in her quiver, ready to annihilate ordered regiments of human artefacts, and their frequently arrogant creators. But the promethean genius of these little and restless inhabitants of planet Earth is that they have not just hidden from the fury of the natural elements, or simply built defences against them, they have harnessed them. Natural forces have been subjugated and bent to the will of and to the yearnings of humanity, or that part of it which has the capacity to decide and command. Here a twofold necessity for knowledge manifests: on one hand is the necessity of protection against natural forces, on the other the necessity of using the same forces for practical purposes. To achieve either of these goals, it is necessary to understand the force one wishes to control. A straightforward example of this dualism is the dam, which protects the valley from a mountain's unpredictable water fluxes, and is also an essential component for the operation of the hydroelectric power plant which transforms the water's force into readily available and non-destructive energy. We also find this essential ambivalence when we consider instability. Our first concern is to preserve the *status quo*. For instance, we have a bridge, a building, or an engine, we do everything in our power to ensure that these artefacts continue to perform their function for as long as is possible. We make sure that they are robust and safe from influences from external elements which may alter their structure, function, or shape. We regularly inspect and maintain them so that possible vulnerabilities are identified and reduced. This is the case in control systems and governors, where the attempt to mitigate the arising of instabilities has been a subject of research since the Industrial Revolution.

Another example of a system where exerting the maximum control is desirable to protect against instabilities is the power grid of a country. The power grid is a complex system constituted by a vast number of nodes, i.e. atomic power plants, hydroelectric power plants, wind farms, small local generators, photovoltaic cells located on the rooves of private houses which feed their excess production back to the grid, large industrial users, and small domestic ones. All these nodes are connected in a network in which energy is produced, travels, and is consumed. To the topological complexity of such networks, the fact that the quantity of energy generated by individual generation units changes over time depends substantially on the heterogeneity of different sources. For instance, photovoltaic cells do not produce energy during the night, or they produce less on cloudy days, while the output of wind turbines depends on wind intensity. In this complex scenario, keeping a stable and constant energy supply is essential to the operation of an advanced economy and to guaranteeing the material wellbeing of modern societies who depend on technology. It becomes necessary that local variations in the production and input of energy, as well as variation in the output required by users, do not create fluctuations capable of destabilising the system, causing *blackouts* and supply interruptions, and bringing production activities and national security to their knees. In order for the system distributing electrical energy to operate in a stable way, it is necessary that all generators oscillate at the same frequency (typically 50 Hertz), that they are synchronised, and, critically, that they can rapidly return to a synchronous oscillation state following perturbations that can be caused by a malfunctioning component or a change in energy supply at a certain point in the network. The reliable preservation of these criteria is influenced by the quantity of electricity supplied to the network by the various sources participating to production at a given moment of time, and by the quantity and consumption of users, which also varies. The general stability of such a complex system is maintained using precise *feedback* systems, which monitor the essential parameters and adjust the power supply accordingly. The optimisation of the stability conditions of a system crucial to millions of people is, obviously, the object of continuous research.

Another relevant example of the control of instability can be found in nuclear fusion reactors, which we have discussed previously using as an example the Rayleigh–Taylor instability, where the stability of plasma must be regulated. In this case, the control of instability results in more efficient and long-lasting energy production, and this can be very useful for humanity. Various instabilities take place in plasmas used in nuclear fusion reactors, and a diverse range of control methods exists. Some of them use, for instance, feedback systems, injection of gas jets, electromagnetic fields, or laser beams with oscillating intensity.

For instance, the possibility of mitigating Rayleigh–Taylor instability in plasmas can be based on the use of a laser beam whose intensity periodically oscillates, causing a variation of plasma acceleration, or on intentionally inducing small oscillating perturbations in the plasma with selected desired features.

In the context of fluids, the understanding of the nature of turbulence phenomena caused by instability processes is necessary in other crucial domains of engineering, for instance the optimal design of airplanes turbines, of combustion engines, as well as in the development of hydraulic systems.

Many practical applications of lasers in technology require their light emission to be stable, regular, and with the least possible sensitivity to the effects of external perturbations. It can be necessary to make sure that the laser oscillation frequency – the "colour" of the emitted light – does not change over time, or that the light beam is homogeneous and does not exhibit fluctuations. For this reason, expedients are taken during the design process in order to minimise undesired effects and which suppress possible instabilities that could otherwise arise. Instability control methods in lasers employ *feedback loop* mechanisms where part of the emitted light is collected outside of the laser, manipulated, and subsequently reinjected or used to generate suitable electronic signals which modify in real time some features of the laser itself, like the optical cavity length or the energy stored in the atoms of its active medium. In a moment we will see how in certain cases, instability processes in lasers are incentivised to reach certain specific goals.

The second way is the one which sees instabilities as a resource, energies that can be made to work for us, channelled to achieve our goals and desires. In these cases one does not attempt to suppress instabilities; on the contrary, one tries to let them maximally develop in appropriate contexts and forms. If we reflect for a moment, we see just how frequently in history political and cultural revolutions have succeeded in situations where the *status quo* already appeared to be about to collapse by exploiting social instability or a predisposition to it. In such cases, there has often been a conscious intent to change the course of events and to reach a desired result, but the energy, the potential for change, the instability was a necessary ingredient if the process of transformation is to succeed. Intentional events would be devoid of efficacy if executed in times which were "not ripe", where a sort of propensity to instability was not already present. It is also possible to consider the instabilities manifesting in natural physical systems or in the behaviour of artefacts and human made devices from this perspective.

A case in which the instability of a natural system is useful for humanity is surely natural radioactivity, and a particular example is the decay process of carbon 14. Carbon 14 is a particular type – or isotope[1] – of carbon, and it is radioactive. It forms in the terrestrial atmosphere in an interaction process between cosmic rays and nitrogen atoms. During their lives, all living organisms absorb it together with carbon 12 and carbon 13 (these are stable and non-radioactive). For plants, the absorption of carbon occurs through photosynthesis. Animals absorb it from other organic substances through the breathing and nutrition processes. In fact, the concentration of these elements present in a living organism during its life corresponds to the atmospheric one. When the organism dies, carbon assimilation ceases. And here comes our friend, instability; carbon 14 atoms contained in the organism, being unstable, start to decay through the beta decay process, producing nitrogen 14 atoms. Approximately 5,730 years are necessary for the quantity of carbon 14 atoms present inside a given organism to halve as a consequence of decay. By measuring the ratio between the quantity of carbon 14 and of carbon 12 (the latter does not decay and remains constant) present in an organism, we can calculate its age. This dating system, discovered by the American Willard Libby in the 1940s, allows archaeologists, historians, anthropologists, and geologists to date fossils and finds containing organic materials like bones, wood, or other animal and plant matter, dating back around fifty thousand years. For instance, the date that mammoths became extinct and the age of the ancient Dead Sea Scrolls have been established using the carbon 14 method. However, the explosion of nuclear devices and the introduction into the atmosphere of great quantities of fossil fuel during the last century have caused a significant alteration of the quantity of carbon 14 present in the atmosphere, which has proved to be a substantial problem for the dating method. For this reason, it will be impossible for future generations to use this method to date finds from after the nineteenth century. Summarising in a single sentence: we can say that carbon 14's instability and decay constitute a sort of telescope which allows us to view a detailed panorama of the past, a galaxy of dates, that could not be seen with the naked eye.

Photonic technologies are another field where the harnessing and exploitation of instabilities in view of useful application is particularly promising. In the collective imagination, the laser is the device which best symbolises and embodies the idea of photonics: the ensemble of those technologies employ light as a computational, precision measurement, material processing, military, and healthcare tool. During the twenty-first century photonics will very likely become a more and more dominant technology in the communication, computation, measurement, and military processes. Many predict that it will have a role comparable to that played by electronics in the twentieth century.

An example where instability plays a decisive role in the operation of lasers is the spontaneous emission process which we have previously seen. The instability of excited atomic states which lead to decay and the spontaneous emission of photons is the mechanism that enables the start of lasers and ultimately their operation. In this sense, such a fundamental process, caused by quantum vacuum fluctuations, is a fundamental ingredient for the existence of many of the technologies that permeate our daily lives, but it does not end there.

One of the most useful features of lasers is their ability to emit a train of light pulses, a repeating alternation of light and darkness. This alternation is very rapid, and not discernible to the naked eye. It occurs at a frequency that can have the order of magnitude of the megahertz (a million times per second) and the gigahertz (a billion times per second). In this very short time the light intensity emitted by the laser goes from zero – darkness – to a peak of extremely intense light, then fades again. This phenomenon repeats indefinitely while the laser remains on. The duration of the pulses varies depending on the kind of laser, but in the most extreme cases it can be equal to a few femtoseconds (a femtosecond is equivalent to a millionth of a billionth of a second). In general, we can say that a laser operates in a continuous wave regime. This means that the generated light consists of a flux whose intensity is constant in time. This continuous flux in time corresponds to a single frequency of the electromagnetic field, to a single colour if we wish, to a single resonance of the optical cavity where the active medium is confined. The regime where the laser emits light pulses is enabled by an instability of the continuous wave emission. The instability brought about by the introduction of peculiar devices to the optical cavity or by a sufficiently large injection of energy into the active medium manifests itself in the amplification of small light waves, initiated by optical noise, which oscillate at different frequencies allowed by the cavity. These waves, oscillating in a synchronous fashion, are amplified and give origin to a spectrum of distinct frequencies. The synchronous oscillation, a sort of cooperation among the waves which arises from constructive interference, ensures that the light radiation circulating inside the laser acquires the form of light pulses. Every time that an impulse hits one of the partially reflecting cavity mirrors, a part of its energy is reflected and remains inside, while another part exits the laser into the external world. In technical jargon, this regime of operation is called *mode-locking*, a term meaning more or less that the different modes of oscillation of the system are "constrained" to move together in a synchronous fashion. Lasers generating light pulses are extremely useful and have many practical applications. One of these is material processing, where light pulses are used like drills and chisels constituted of pure energy. Furthermore, many surgical operations in ophthalmology – connected to cataract removal, myopia, and other sight defects correction – are performed using lasers that emit ultrashort pulses with a duration approximately equal to a femtosecond. These pulses are precisely focussed on the eye area requiring intervention and act like a scalpel of light, selectively incising and cutting the necessary areas without the risk of damaging others. At a conference Gérard Mourou, pioneer of the ultrashort high-energy laser pulses generation technique for which he was awarded the Nobel price together with Donna Strickland, told us how the discovery of this beneficial application was not down to an intelligent, rational, and well-informed study, or due to well-intentioned reasoning, nor was it the result of a well devised experiment. He told us that one day, one of his students was working in the lab without wearing protective goggles and was hit in the eye by a powerful laser beam. The physician who visited him was amazed by the perfect, delimited, and precise damage to his eye. From this episode imbued by serendipity the idea was born of using these lasers for eye surgeries.

A further important practical use of instabilities in the field of photonic technologies is the one constituted by the so-called optical frequency combs generated in optical resonators and in pulsed laser. Optical frequency combs are states of light consisting of large numbers of electromagnetic waves (colours) that are equally spaced in frequency and oscillate in a synchronous fashion. A frequency comb constitutes a sort of optical ruler, i.e. an ultraprecise tool allowing the measurement of distances equal to the dimensions typical of an atom, or at time intervals with a precision superior to an atomic clock. Indeed, it is most likely that the caesium atom clocks which, at the moment, provide the universal standard for the measurement of time will be superseded by optical clocks operating thanks to optical frequency combs.

Let us take a closer look at the formation process of these particular states of light in the case of optical resonators. Previously, in the context of optical Turing instability, we have seen how if a single light wave oscillating at a certain frequency is injected into an optical resonator, it can become unstable and cede energy to other waves resonating inside the cavity. This process initially generates two waves, one with a lower and one with a higher frequency. But the process repeats as a cascade, giving origin to an extremely large number of frequencies, in certain cases a few thousands. Under specific conditions these waves oscillate in a perfectly synchronous way and their frequencies are equally spaced. They constitute what is called a frequency comb, which covers a multitude of different colours even sometimes extending into regions of the electromagnetic spectrum which are not visible to the human eye, such as the near and mid infrared. Frequency combs are used as precision sensors to determine which chemical substances are present in a material or to measure the concentration of greenhouse gasses in the atmosphere. In these cases, light emitted by the optical resonator is shone on the material which one wants to analyse and then the reflected or transmitted light is collected with suitable detectors. The light which has interacted with the material, or with the atmosphere, carries imprints of traces of the substances that are contained within it, which can then be communicated to the researcher. What is fascinating about this peculiar kind light source is the possibility of creating them with miniaturised dimensions, for instance an optical resonator with a diameter of less than one millimetre. The use of these sources opens new and fascinating horizons for a multitude of applications, in an absolutely non-invasive fashion, on portable devices or in contexts where the necessity of minimizing weight and size is vital, as is the case for drones, satellites, or rockets. In this respect, a case of formidable interest is the use of frequency combs for space exploration and in particular for searching for exoplanets, planets with features similar to the ones of Earth orbiting stars in other solar systems.

Also in the field of optical devices, engineering the instabilities of light waves, and in particular modulation instability, is at the heart of current research into so-called *parametric amplifiers*, which are excellent candidates to increase the amount of information that can be transmitted in parallel in telecommunication systems. The backbone of the modern physical infrastructure of the internet is constituted by a dense network of optical fibres, ever expanding across continents and on ocean floors. Most of the data traffic associated with the internet travels in the form of light waves and impulses through these optical fibres. However, the fibres are imperfect and energy losses take place while signals travel. This fact, which leads to a degradation of the transmitted information, makes it necessary that signals are regenerated by periodically giving them new energy during their journey. This is done using devices called optical amplifiers. There are, in reality, various effective ways to implement the amplification. However, the ever-increasing demand for greater quantities of transmitted information by modern interconnected societies means that more and more signals, each one corresponding to a different frequency, must be simultaneously reenergised. Here, modulation instability, with its ability to transfer energy from a powerful wave to many small wavelets oscillating at different frequencies, comes into play. In this case, optical signals with low power carrying information are injected inside a special fibre together with a powerful light wave called a pump, and act as perturbations to the latter. The instability of the pump wave translates, under suitable conditions, into the energy transfer to signals carrying information, which eventually exit, amplified, from the fibre. Optical amplifiers operating thanks to this principle are currently the object of active research.

The examples presented in this chapter are only a sample, which should give an idea of the practical importance of knowing and characterising with precision instability processes. They also show how the path of harnessing energies and the latent possibilities offered by instabilities is variegated and promising, and how it stretches across many disciplines and offers each of them a precious resource. In every problem, a gift.

During my days of research, while I try to find meaning in the mathematical equations in front of my eyes, or when I discuss recent experimental measurements obtained in the laboratory with my collaborators, the thought of the twofold nature of instability, and in particular of how to transform it into a resource, is sometimes pleasing one. At others it is an obsession. It is untameable, and resurfaces in unexpected moments: while I cycle in the countryside; in front of a pint of beer in the late afternoon crowd of a busy pub; in the lazy evening idleness, as if it would ask me to explain how I use my time…

NOTE

1. An isotope of a chemical element possesses in its nucleus the same number of protons that characterise that element, but differs in the number of neutrons. Different carbon isotopes exist, for instance carbon 14 possess 6 protons and 8 neutrons whose total – also called the mass number – gives 14; instead carbon 12 possess 6 protons and 6 neutrons, and hence it has mass number equal to 12.

CHAPTER 17

Becoming

Listen, O lord of the meeting rivers, things standing shall fall, but the moving ever shall stay.

BASAVA

In 1902 a scientific paper was published in which the British physicist Sir James Jeans proved that when, in a cloud of gaseous matter, the density exceeds a certain critical value, pressure is no longer capable of opposing the process that leads the gas to gather infinitely, and to eventually collapse into a single point. What would be later called the *Jeans' instability* is the mechanism at the basis of the formation of stars and galaxies, which begin with the spontaneous collapse of enormous clouds of gas and dust called nebulae. This birth is a counterbalance to the catastrophically destructive role that instability can play at a cosmic and planetary level, and with which our journey began. I like to imagine how cycles of creation and destruction follow one another; how, from fluctuating dusts, entire solar systems, new stars can arise in the firmament, but which, with a transience that can be understood only from the viewpoint of billions of lifetimes, dissolve again, only to be reborn once more, as if in Eternal Return. Birth and Death animated by instability, change as the sole constant.

DOI: 10.1201/9781003462040-17

A fusion of Nietzsche and Eraclitus that, while one hand closes the narrative circle of these pages, with the other it describes, in natural phenomena, the opening of the present simultaneously towards the unfolding of the future, and towards the interpretation of the past.

In the initial chapters we hinted at the power of language, at the form that it gives to our experience of the world. The concept of instability, and the way in which it has been developed in modern sciences, and in particular in physics, in its various declinations, is a powerful interpretative key which unlocks the understanding of a vast class of phenomena in cosmic spaces, in the heart of atoms, in chemical reactions, on waves in a fluid, and between the mirrors of a laser. Some of these phenomena could probably be explained differently in the context of other interpretation paradigms, or they could appear to be of a different nature if seen with different eyes, with eyes which do not belong to mathematical science, for instance under the gaze of poetry, of mysticism, or of philosophy. The fact is that the idea of instability, as it has been developed by modern science, has proved itself capable of shedding light on the qualitative and revolutionary processes of change which happen in multiple complex systems, irrespective of the constitution of the material substrate of the latter. The concept of instability is intrinsic to processes occurring both suddenly and in an exponential fashion, to abrupt transitions that completely transform the landscape, to chaos, and to the creation of new forms of spontaneous and self-organised order.

From an existential point of view, the human being, in all historical epochs – and our own epoch of technique is no exception – has always sought to take advantage of tools and narratives that can help to predict the future before it becomes the actual, and to justify it once it has manifested, becoming the present and then the past. In different forms, religion and myth have tried to fulfil this task in past epochs. In the current one, the structural and seismic changes which society and nature undergo in the face of the impetus of technological developments and of the crumbling of the traditional social order do not relieve us from questions about the direction in which the world will go, about what awaits us, about the forms of becoming. On the contrary, these rapid and dramatic movements exacerbate our anxiety and fear. New tools become necessary to face new uncertainty, and in the thought paradigms that have arisen in the last centuries, in the scientific environment significant resources can possibly, by analogy, be found. The concept of instability with its equilibrium situations, with its perturbations and their exponential growth, with its bifurcations and transformative and self-organised changes, can at least qualitatively offer a modest foothold. As an intellectual exercise, we can learn to observe the processes and the events which manifest themselves around us, at the social, cultural, political, and environmental level, as well as in our personal, work, and relational dynamics through the prism of instability; we can try tracing the constitutive ideas we have explored on these pages. Possibly, this effort will lead us to the appropriate new categories of the conceptualisation of becoming, and to immerse ourselves in different hermeneutics of the ordinary, to broaden our meaning horizon, to foresee processes in essence still weaved in the fog of the non-manifest, to be less slaves to a horizontal present. If this will not serve to predict the changes, it will perhaps make their acceptance less tragic and their interpretation more fulfilling.

Ours is an epoch where novelty constantly peeks out through the grainy meshes of the fabric of the world, manifesting through different social and technological forms, through different forms of power management, of relation among human beings and between the human and nature. During restless and turbulent times, the form of interpretative thought offered by instability theory is, more than ever, actual and alive as an instrument that, even with all its limits, accounts for the dearth and impermanence of the *status quo*, and at the same time sheds a faint light on the becoming which devours it.

Bibliography

PRELUDE IN ORDINARY LANGUAGE

Wittgenstein L., *Philosophical Investigations*, Wiley-Blackwell, Chichester, 2009.

PLANETARY INSTABILITIES

Barrow-Green J., *Poincaré and the Three Body Problem*, American Mathematical Society, 1997.

Crowe M.J., *Theories of the World from Antiquity to the Copernican Revolution*, Dover Publications, New York, 1990.

Fitzpatrick R., *An Introduction to Celestial Mechanics*, Cambridge University Press, Cambridge, 2012.

Galilei G., *Dialogue Concerning the Two Chief World Systems*, University of California Press, Berkeley, 1953.

Goldstein H., Poole C., Safko J., *Classical Mechanics*, 3rd edition, Addison Wesley, San Francisco, 2000.

Laskar J., Is the Solar System Stable? In Duplantier B., Nonnenmacher S., Rivasseau V. (Eds.) *Chaos. Progress in Mathematical Physics*, vol. 66. Birkhäuser, Basel, pp. 239–270, 2013.

Maxwell J.C., On the Stability of the Motion of Saturn's Rings. In W. Niven (Ed.) *The Scientific Papers of James Clerk Maxwell* (Cambridge Library Collection - Physical Sciences). Cambridge University Press, Cambridge, pp. 288–376, 1855 [2011].

Speiser D., The Kepler Problem from Newton to Johann Bernoulli. *Archive for History of Exact Sciences*, vol. 50, pp. 103–116, 1996.

CYBERNETICS AND CONTROL SYSTEMS

Anderson P.M., Fouad A.A., *Power system control and stability*, Wiley-IEEE Press, Piscataway, 2003.

Ashby W.R., *An Introduction to Cybernetics*, Chapman & Hall, London, 1956.

Kang C.-G., Origin of Stability Analysis: 'On Governors' by J.C. Maxwell. *IEEE Control Systems Magazine*, vol. 36, pp. 77–88, 2016.

Maxwell J.C., On Governors. *Proceedings of the Royal Society of London*, vol. 16, pp. 270–283, 1868.

HYDRODYNAMICS INSTABILITIES: CHAOS AND TURBULENCE

Chandrasekhar S., *Hydrodynamic and Hydromagnetic Stability*, Dover Publication, New York, 1981.

Craik A.D.D., *Wave Interactions and Fluid Flows*, Cambridge University Press, Cambridge, 1985.

Faraday M., On a Peculiar Class of Acoustical Figures; and on Certain Forms Assumed by a Group of Particles Upon Vibrating Elastic Surfaces. *Philosophical Transactions of the Royal Society (London)*, vol. 121, pp. 299–340, 1831.

Strecker K.E. et al., Formation and Propagation of Matter-Wave Soliton Trains. *Nature*, vol. 417, pp. 150–153, 2002.

MODULATION INSTABILITY

Benjamin T.B., Instability of Periodic Wavetrains in Nonlinear Dispersive Systems. *Proceedings of the Royal Society of London A*, vol. 299, pp. 59–75, 1967.

Bogolyubov N.N., On the Theory of Superfluidity. *The Journal of Physics, USSR*, vol. 11, pp. 23-32, 1947.

Mussot A. et al., Fibre Multi-Wave Mixing Combs Reveal the Broken Symmetry of Fermi-Pasta-Ulam Recurrence. *Nature Photonics*, vol. 12, pp. 303–308, 2018.

Pierangeli D. et al., Observation of Fermi-Pasta-Ulam-Tsingou Recurrence and Its Exact Dynamics. *Physical Review X*, vol. 8, p. 041017, 2018.

Tai K., Hasegawa A., Tomita A., Observation of Modulational Instability in Optical Fibers. *Physical Review Letters*, vol. 56, pp. 135–138, 1986.

Zakharov V.E., Ostrovsky L.A., Modulation Instability: The Beginning. *Physica D: Nonlinear Phenomena*, vol. 238, pp. 540–548, 2009.

AN ACT OF CREATION

Dauxois T., Peyrand M., *Physics of Solitons*, Cambridge University Press, Cambridge 2006.

Dudley J.M. et al., Rogue Waves and Analogies in Optics and Oceanography. *Nature Reviews Physics*, vol. 1, pp. 675–689, 2019.

Mollenauer L.F., Gordon J.P., *Solitons in Optical Fibers: Fundamentals and Applications*, Elsevir Academic Press, Burlington, 2006.

Novikov S., Manakov S.V., Pitaevskii L.P., Zakharov V.E., *Theory of Solitons: The Inverse Scattering Method*, Plenum, New York, 1984.

Onorato M., Residori S., Baronio F. Editors, *Rogue and Shock Waves in Nonlinear Dispersive Media*, Springer, Berlin Heidelberg, 2016.

Russel J.S., *Report on Waves*, "Report of the fourteenth meeting of the British Association for the Advancement of Science held at York in September 1844", Plates XLVII–LVII, John Murray, London, pp. 311–390, 1845.

Trillo S., Torruellas W. (Editors), *Spatial Solitons*, Springer-Verlag, Berlin Heidelberg, 2001.

Zakharov V.E., Dyachenko A.I., Prokofiev A.O., Freak Waves as Nonlinear Stage of Stokes Wave Modulation Instability. *European Journal of Mecahnics - B/Fluids*, vol. 25, pp. 677–692, 2006.

BIFURCATIONS

Arnold V.I., *Catastrophe Theory*, 3rd edition, Springer-Verlag, Berlin Heidelberg, 1992.

Lugiato L., Prati F., Brambilla M., *Nonlinear Optical Systems*, Cambridge University Press, Cambridge, 2015.

Strogatz S., *Nonlinear Dynamics and Chaos: With Applications to Physics, Biology, Chemistry, and Engineering*, 2nd edition, CRC Press, Boca Raton, 2015.

Svelto O., *Principles of Lasers*, Springer, New York, 2010.

MORPHOGENESIS

Copeland B.J. et al., *The Turing Guide*, Oxford University Press, Oxford, 2017.

Cross M.C., Hohenberg P.C., Pattern Formation Outside of Equilibrium. *Reviews of Modern Physics*, vol. 65, pp. 851–1112, 1993.

Getzin S. et al., Bridging Ecology and Physics: Australian Fairy Circles Regenerate Following Model Assumptions on Ecohydrological Feedbacks. *Journal of Ecology*, 2020. DOI: 10.1111/1365-2745.13493.

Gierer A., Meinhardt H., A Theory of Biological Pattern Formation. *Kybernetik*, vol. 12, pp. 30–39, 1972.

Haelterman M., Trillo S., Wabnitz S., Dissipative Modulation Instability in a Nonlinear Dispersive Ring Cavity. *Optics Communications*, vol. 91, pp. 401–407, 1992.

Lejeune O., Tlidi M., Couteron P., Localized Vegetation Patches: A Self-Organized Response to Resource Scarcity. *Physical Review E*, vol. 66, p. 010901(R), 2002.

Lugiato L., Prati F., Brambilla M., *Nonlinear Optical Systems*, Cambridge University Press, Cambridge, 2015.

Nicolis G., *Introduction to Nonlinear Science*, Cambridge University Press, Cambridge, 1995.

Prigogine I., Stengers I., *La nuova alleanza. Metamorfosi della Scienza*, Einaudi Editore, Torino, 1999.

Prigogine I., Glansdorff P., *Thermodynamic Theory of Structure, Stability and Fluctuations*, Wiley-Interscience, London, 1971.

Tlidi M., Lefever R., Vladimirov A., On Vegetation Clustering, Localized Bare Soil Spots and Fairy Circles, chapter in Akhmediev N., Ankiewicz A. (Eds.) *Dissipative Solitons: From Optics to Biology and Medicine*, Lec. Notes Phys. 751, Springer, Berlin Heidelberg, 2008.

Tripaldi L., *Parallel Minds: Discovering the Intelligence of Materials*, Urbanomic, Falmouth, 2022.

Turing A., The Chemical Basis of Morphogenesis. *Philosophical Transactions of the Royal Society of London B*, vol. 237, pp. 37–72, 1952.

INSTABILITIES AT THE ATOMIC SCALE

Povh B. et al., *Particles and Nuclei*, Springer-Verlag, Berlin Heidelberg, 2006.

Griffiths D., *Introduction to Elementary Particles*, Wiley-VCH Verlag, Weinheim, 2008.

QUANTUM VACUUM AND THE ORIGIN OF THE PERTURBATIONS

Aylesworth G., Postmodernism. In *The Stanford Encyclopedia of Philosophy (Spring 2015 Edition)*, Edward N. Zalta (Eds.)

URL = <https://plato.stanford.edu/archives/spr2015/entries/postmodernism/>

Einstein A., Zur Quantentheorie der Strahlung. *Physikalische Zeitschrift* 18, pp. 121–128, 1917.

Griffiths D.J., *Introduction to Quantum Mechanics*, Pearson Prentice Hall, Upper Saddle River, 2005.

Meister Eckhart, *The Complete Mystical Works of Meister Eckhart*, The Crossroad Publishing Company, New York, 2009.

Milonni P.W., *The Quantum Vacuum: An Introduction to Quantum Electrodynamics*, Academic Press, San Diego 1994.

Rovelli C., *Helgoland*, Penguin Books, 2022.

Scully M.O., Zubairy M.S., *Quantum Optics*, Cambridge University Press, Cambridge, 1997.

Srednicki M., *Quantum Field Theory*, Cambridge University Press, Cambridge, 2007.

Suzuki D.T., *Mysticism – Christian and Buddhist – The Eastern and Western Way*, Harper & Row, New York, 1971.

Vattimo G., *The End of Modernity*, Polity Press, Cambridge, 1988.

INSTABILITIES IN THE HEAD

Izhikevich E.M., *Dynamical Systems in Neuroscience: The Geometry of Excitability and Bursting*, MIT Press, Cambridge, 2006.

Prucnal P.R., Shastri B.J., *Neuromorphic Photonics*, CRC Press, Boca Raton, 2017.

THE PILLARS OF HERCULES

Dauxois T., Fermi, Pasta, Ulam, and a Mysterious Lady. *Physics Today*, vol. 61, pp. 55–57, 2008.

Fermi E., Pasta J., Ulam S., *Studies of Nonlinear Problems*, Document LA-1940. Los Alamos National Laboratory, 1955.

Strogatz S., *Nonlinear Dynamics and Chaos: With Applications to Physics, Biology, Chemistry, and Engineering*, 2nd edition CRC Press, Boca Raton, 2015.

EXPONENTIAL NATURE

Kurzweil R., *The Age of Spiritual Machines: When Computers Exceed Human Intelligence*. Penguin Books, New York, 1999.

Kurzweil R., *The Law of Accelerating Returns*. https://www.kurzweilai.net/the-law-of-accelerating-returns, 2001.

UNIVERSAL FORMS

Ablowitz M.J., *Nonlinear Dispersive Waves*, Cambridge University Press, Cambridge, 2011.

Aranson I.S., Kramer L., The World of the Complex Ginzburg-Landau Equation. *Review of Modern Physics*, vol. 74, pp. 99–143, 2002.

Fedele R., From Maxwell's Theory of Saturn's Rings to the Negative Mass Instability. *Philosophical Transactions of the Royal Society A*, vol. 366, pp. 1717–1733, 2008.

Haken H., *Synergetics: Introduction and Advanced Topics*, Springer-Verlag, Berlin Heidelberg, 2004.

Wigner E., The Unreasonable Effectiveness of Mathematics in the Natural Sciences. Richard Courant lecture in mathematical sciences delivered at New York University, May 11, 1959. *Communications on Pure and Applied Mathematics*, vol. 13, pp. 1–14, 1960.

RIDING THE TIGER

Andrekson P.A., Karlsson M., Fiber-Based Phase-Sensitive Optical Amplifiers and their Applications. *Advances in Optics and Photonics*, vol. 12, pp. 367–428, 2020.

Cundiff S.T., Ye J., Colloquium: Femtosecond Optical Frequency Combs. *Review of Modern Physics*, vol. 75, pp. 325–342, 2003.

Fortier T., Baumann E., 20 Years of Developments in Optical Frequency Comb Technology and Applications. *Communications Physics*, vol. 2, p. 153, 2019.

Gajduk A., Todorovski M., Kocarev L., Stability of Power Grids: An Overview. *The European Physical Journal Special Topics*, vol. 223, pp. 2387–2409, 2014.

Haus H., Mode-Locking of Lasers. *IEEE Journal on Selected Topics in Quantum Electronics*, vol. 6, pp. 1173–1185, 2000.

Hender T.C. et al., MHD Stability, Operational Limits and Disruptions. *Nuclear Fusion*, vol. 47, pp. S128–S202, 2007.

Kawata S., Karino T., Gu Y., Dynamic Stabilization of Plasma Instability. *High Power Laser Science and Engineering*, vol. 7, p. E3, 2019. DOI: 10.1017/hpl.2018.61.

Marra J.F., *Hot Carbon: Carbon 14 and a Revolution in Science*, Columbia University Press, New York, 2019.

Pasquazi A. et al., Micro-Combs: A Novel Generation of Optical Sources. *Physics Reports*, vol. 729, pp. 1–81, 2018.

BECOMING

Falco M., Hansen S.H., Wojtak R., Mamon G.A., Why Does the Jeans Swindle Work? *Monthly Notices of the Royal Astronomical Society*, vol. 431, pp. L6–L9, 2013.

Jeans J.H., The Stability of a Spherical Nebula. *Philosophical Transactions of the Royal Society of London. Series A, Containing Papers of a Mathematical or Physical Character*, vol. 199, pp. 1–53, 1902.

Index

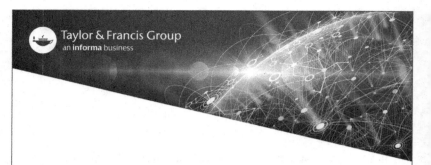

Printed in the United States
by Baker & Taylor Publisher Services